Environmental Science

FOR

DUMMIES®

Environmental Science

FOR

DUMMIES®

by Alecia M. Spooner

John Wiley & Sons, Inc.

Environmental Science For Dummies®

Published by
John Wiley & Sons, Inc.
111 River St.
Hoboken, NJ 07030-5774
www.wiley.com

Copyright © 2012 by John Wiley & Sons, Inc., Hoboken, New Jersey

Published simultaneously in Canada

For general information on our other products and services, please contact our Customer Care Department within the U.S. at 877-762-2974, outside the U.S. at 317-572-3993, or fax 317-572-4002.

For technical support, please visit www.wiley.com/techsupport.

Wiley publishes in a variety of print and electronic formats and by print-on-demand. Some material included with standard print versions of this book may not be included in e-books or in print-on-demand. If this book refers to media such as a CD or DVD that is not included in the version you purchased, you may download this material at http://booksupport.wiley.com. For more information about Wiley products, visit www.wiley.com.

Library of Congress Control Number: 2012941753

ISBN 978-1-118-16714-4 (pbk); ISBN 978-1-118-22630-8 (ebk); ISBN 978-1-118-23961-2 (ebk); ISBN 978-1-118-26421-8 (ebk)

Manufactured in the United States of America

10 9 8 7 6 5 4 3 2 1

WILEY

About the Author

Alecia M. Spooner teaches Earth and environmental sciences at Everett Community College in Everett, Washington. She has degrees in anthropology (a BA from the University of Mississippi), archaeology (an MA from Washington State University), and geology (an MS from the University of Washington). In graduate school, she reconstructed paleoclimate and paleoenvironments from fossil pollen records in lake sediments of the Pacific Northwest. She teaches by using active learning and inquiry. She lives in Shoreline, Washington, with her husband, two boys, a cat, and two chickens.

Dedication

To the many teachers who inspired me to look at the world more closely.

"We shall never achieve harmony with land, any more than we shall achieve absolute justice or liberty for people. In these higher aspirations, the important thing is not to achieve but to strive."

—Aldo Leopold

Author's Acknowledgments

This book was created with support and encouragement from the following people: Igor, Pasha and Dima, Mom and Dad, Heather and Julie, Alli and Holly, Janice and Kysa, Rene and Matt, and Elizabeth, Amanda, and Lindsay. Thank you.

Publisher's Acknowledgments

We're proud of this book; please send us your comments at http://dummies.custhelp.com. For other comments, please contact our Customer Care Department within the U.S. at 877-762-2974, outside the U.S. at 317-572-3993, or fax 317-572-4002.

Some of the people who helped bring this book to market include the following:

Acquisitions, Editorial, and Vertical Websites

Project Editor: Elizabeth Rea

Executive Editor: Lindsay Sandman Lefevere

Copy Editor: Amanda M. Langferman

Assistant Editor: David Lutton

Editorial Program Coordinator: Joe Niesen

Technical Editor: J. Daniel Husband, Ph.D.

Editorial Manager: Michelle Hacker

Editorial Assistant: Alexa Koschier

Art Coordinator: Alicia B. South

Cover Photos: © iStockphoto.com/Fred Froese

Cartoons: Rich Tennant (www.the5thwave.com)

Composition Services

Project Coordinator: Patrick Redmond

Layout and Graphics: Carl Byers, Carrie A. Cesavice, Joyce Haughey, Christin Swinford

Proofreaders: John Greenough, Tricia Liebig

Indexer: Sharon Shock

Illustrator: Lisa Reed

Publishing and Editorial for Consumer Dummies

    Kathleen Nebenhaus, Vice President and Executive Publisher

    Kristin Ferguson-Wagstaffe, Product Development Director

    Ensley Eikenburg, Associate Publisher, Travel

    Kelly Regan, Editorial Director, Travel

Publishing for Technology Dummies

    Andy Cummings, Vice President and Publisher

Composition Services

    Debbie Stailey, Director of Composition Services

Contents at a Glance

Table of Contents

Introduction

*E*nvironmental science is the study of Earth's environment. To study the environment, scientists draw from many other disciplines, including chemistry, geography, economics, and everything in between. No wonder students new to environmental science often find themselves dizzy at the breadth of information needed to study and solve environmental problems.

Fortunately, you've found *Environmental Science For Dummies*! Regardless of whether you picked up this book to help you through a science class or to begin an independent exploration of environmental science, I hope it's a useful reference for you, providing an introduction to the most important concepts and issues in modern environmental science.

I've written this book to cover as many environmental science concepts as possible, while at the same time helping you understand how these concepts apply to your life. If you're already familiar with some of the topics explained in the book, perhaps this book will help renew your interest in environmental studies and sustainable living. If these topics are completely new to you, I hope it helps you realize that you can take action daily and make choices that affect your environment in a positive way.

About This Book

Environmental Science For Dummies presents an introduction to the core concepts in environmental science and the most important issues studied by environmental scientists today.

The topics in environmental science are so intricately linked that there's simply no way to explain one without explaining a little bit of another as well. In each chapter, I use cross-references to other chapters to help you link together the related concepts and to provide a more complete understanding of the complex topics in environmental science.

Throughout the book, you also find multiple illustrations. These drawings expand on what I've written in places where a visual representation may be helpful. But don't forget to look up from reading once in a while! You have plenty of first-hand experience with the environment. As you read about certain topics or issues, you may find it useful to look at them in the context of your own life.

Conventions Used in This Book

Here are some of the conventions I use in the book to keep things easy to find and follow:

- ✔ Anytime I use a word that I think you may not have seen before, I put it in *italics* and define it.
- ✔ **Boldface** words highlight a bulleted list or sequence of steps.
- ✔ Internet and web addresses appear in `monotype` to help them stand out.

What You're Not to Read

Throughout this book, you find sidebars highlighted in gray. The sidebars include extra information or particularly interesting tidbits that I thought you might enjoy. I find them interesting — and I hope you do, too — but they aren't required reading to understand the concepts in the book. Feel free to skip these sidebars, as they're not integral to the information presented in each chapter.

Similarly, any portion of text with the Technical Stuff icon beside it indicates that it explains or describes a concept in extra detail, beyond what you need to have a basic grasp of the idea. Feel free to skip these portions or to breeze through them.

Foolish Assumptions

As the author of this book, I've made some assumptions about you, my reader. For instance, I assume that you live on Earth, drink water, breathe air, and use energy for various things such as heating and transportation. I assume that you're familiar with basic geography, such as the location of continents and some countries around the world.

However, I don't assume that you have any background in chemistry, biology, geology, ecology, economics, or any of the other disciplines that are part of environmental science. And you don't need a background in any of these to benefit from the explanations in this book. Wherever the details of another science are important, I provide those details in my explanations.

Each topic in environmental science could fill an entire book of its own, so if you find that something in particular catches your interest, I encourage you to look for books that offer more detail into that topic specifically.

How This Book Is Organized

I've broken this book into chapters and organized those chapters into parts that group topics together. Here's a brief overview of each part.

Part I: Demystifying Science and the Environment

In Part I, I introduce you to environmental science, the study of Earth's environment and the living and nonliving things within it. I describe the scientific method and explain how scientists design effective experiments and portray information by using graphs.

This part includes a discussion of matter, the "stuff" that makes up all things, and a quick look at how atoms bond to form molecules. It describes inorganic matter and the important organic molecules that are the building blocks for life. And it includes a chapter on what scientists understand about energy: what it is, how it works, and how it flows. This is where you find details on photosynthesis and cellular respiration.

Part II: Planting the Seed: Foundational Concepts in Environmental Science

Like any science, environmental science has a few key concepts or principles that provide the foundation for greater understanding. In Part II, I introduce you to these concepts, including how to measure human impact with the ecological footprint and how to use the ecosystem as a unit of study.

The plants and animals that inhabit an ecosystem are determined largely by the climate (temperature and moisture) conditions of the region. Scientists link living communities to climate by classifying ecosystems into defined categories called *biomes,* which I describe in this part.

This part also scratches the surface of *population biology,* which is the study of how organisms interact with one another. I explain competition, cooperation, and predation within an ecosystem, as well as some of the complex ways that scientists measure and track changes in populations (including human populations) over time.

Part III: Getting Your Needs Met: Earth's Natural Resources

A major focus of environmental science is how to use and care for Earth's natural resources so that they can continue to meet the needs of human beings for as long as possible. Part III describes Earth's natural resources and the issues humans face in trying to *conserve* them, or make them last.

You're familiar with some of these resources — water, land, and energy. But you may not realize that the diversity of biological organisms, or *biodiversity,* is also a natural resource. I describe all these resources in this part and explain why biodiversity is so important and why it's in danger in many parts of the world. I also explain the pros and cons of alternative energy sources.

Part IV: Giving a Hoot: Pollution and Environmental Quality

Along with managing natural resources, environmental scientists are often asked to help solve problems created by pollution. Part IV covers topics of environmental quality, including air and water pollution. It's also the place to look for information about what dangerous substances or toxins are present in the environment and how garbage and hazardous waste can be managed to reduce further environmental damage.

This part also addresses what scientists currently understand about Earth's climate and how human actions continue to affect the global climate system.

Part V: Follow the Recycled Brick Road: A Sustainable Future

The goal of most environmental scientists is *sustainability.* This means using the environment and its resources in such a way that it can continue to provide for human needs long into the future, possibly forever. In this part, I describe some basic economic principles and explain how shifting your perspective from human-centered to ecosystem-centered may lead you to make more sustainable choices as a consumer.

Millions of people with different desires and priorities share Earth. In this part, you uncover some of the most successful policies across the U.S. and across the world that have come about to help protect the environment and conserve Earth's resources for future generations.

Part VI: The Part of Tens

In the final part of *Environmental Science For Dummies,* you find three lists. The first is a list of ten simple ways to live life more sustainably. The second describes ten examples of how unsustainable practices have ruined shared resources, or *commons*. And the last chapter of the book lists ten careers that center on environmental science. You may be surprised at the variety of options you have for working in an environmental science field!

Icons Used in This Book

Throughout this book I use icons to catch your eye and highlight certain kinds of information. Here's what these little pictures mean:

Anytime you see the Remember icon take notice! I use this icon to highlight important information, often fundamental to the concepts being explained around it or in the same chapter. Other times I use it to highlight a statement meant to help you pull multiple concepts together.

The Tip icon marks information that may be particularly useful to help you study or prepare for an exam. It often marks a helpful way to remember a certain concept.

The Case Study icon brings your attention to real-world examples of particular environmental issues. Case studies are a great way to provide context for the concepts I present in the book.

Anywhere I describe a potential solution for an environmental problem I mark it with a Solution icon.

A few places in this book I offer a little extra detail about a particular topic or concept and mark it with the Technical Stuff icon.

Where to Go from Here

I've written this book to function as a reference that you can open to any page and dive into. If you choose to start from the beginning, you'll find the information organized in what I hope is a logical way that answers your questions as soon as you think to ask them! But you can also browse the table of contents to find topics you're interested in knowing more about and then turn to the chapters on those topics.

If you've never thought much about how you're connected to everything around you, you may want to start with Chapter 6, which explains what an ecosystem is and does. This chapter may dramatically change your perspective!

If you're intrigued by the idea of alternative energy sources, flip to Chapter 14, where I cover many different ways to fuel daily living without using fossil fuels (coal, gas, and oil) or nuclear power. Environmental scientists have found ways to capture or produce energy in cleaner, more efficient ways than have ever been possible before.

For a real wake-up call, turn to Chapter 18 to see how the packaging and convenience of modern life (think bottles of water, to-go containers, and plastic utensils) have resulted in oceans full of trash. In particular, plastic bits that don't decompose are interfering with ocean ecosystems, which is just one of the consequences of waste I describe in that chapter.

Part I
Demystifying Science and the Environment

The 5th Wave   By Rich Tennant

"You've almost got the Earth's layers right, but I don't think there's a creamy nougat between the mantle and the core."

In this part . . .

At its core, environmental science is like any science — based on a methodical way of asking and answering questions to expand the human understanding of the natural world.

In this part, I describe how the scientific method shapes the process of learning about the environment. I also cover foundational scientific ideas about what makes up everything around you (atoms, molecules, and compounds) and how energy moves things through the environment. This is also where you find out how green plants capture energy from the sun and transform it into sugar through the process of photosynthesis.

Chapter 1

Investigating the Environment

*I*n its simplest terms, *environmental science* is the study of the air you breathe, the water you drink, and the food you eat. But environmental scientists study so much of the natural world and the way humans interact with it that their studies spill over into many other fields. Whether you're a student in a college course or someone who picked up this book to find out what environmental science is all about, you'll find that the ideas in this book apply to your life.

Like any living creature, you depend on environmental resources. More importantly perhaps is the fact that humans, unlike other living creatures, have the ability to damage these resources with pollution and overuse. This chapter provides a quick overview of the environment, its systems, and its many resources. It also talks about what humans can do to reduce their impact on the environment today and into the future. After all, maintaining the health of the Earth and its resources at both the local and global level is something everyone has a stake in.

Putting the "Science" in Environmental Science

Environmental science draws on knowledge from many different fields of study, including the so-called hard sciences like chemistry, biology, and geology and the social sciences like economics, geography, and political science. This section offers a quick overview of some of the scientific concepts, such as how to apply the scientific method to answer questions, that you need to be familiar with as you start your exploration of environmental science. I explain these foundational scientific concepts in more detail throughout the rest of Part I.

Using the scientific method

The *scientific method* is simply a methodical approach to asking questions and collecting information to answer those questions. Although many classes teach it as something that only scientists use, you use it just about every day, too.

You may not write down each step of the scientific method when you use it, but anytime you ask a question and use your senses to answer it, you're using the scientific method. For example, when standing at a crosswalk, you look both ways to determine whether a car is coming and whether an approaching car is going slow enough for you to safely cross the street before it arrives. In this example, you have made an observation, collected information, and based a decision on that information — just like a scientist!

The power of the scientific method is in the way scientists use it to organize questions and answers. It helps them keep track of what's known and what's unknown as they gather more knowledge. This organization becomes particularly important when they study large, complex systems like those found in the natural world. Scientists always have more to learn about the natural world, and using the scientific method is one way that they can follow the path of scientific investigation from one truth to another. Turn to Chapter 2 for more on the scientific method.

Understanding the connection between atoms, energy, and life

Studying the environment includes studying how matter, energy, and living things interact. This is where other fields of study, such as chemistry, physics, and biology, come into play. Here are just a few of the core ideas from these sciences that you need to understand as you study environmental science:

✔ All matter is made of atoms.

✔ Matter is never created or destroyed, but it does change form.

✔ Living matter, or life, is made up of complex combinations of carbon, hydrogen, and oxygen atoms.

✔ Most of the energy at Earth's surface comes from the sun.

✔ Energy transfers from one form to another.

✔ Living things, or organisms, either capture the sun's energy (through *photosynthesis*) or get their energy by eating other living things.

Analyzing the Earth's Physical Systems and Ecosystems

The environment consists of many different systems that interact with one another on various levels. Some systems are physical, such as the *hydrologic system* that transfers water between the atmosphere and the Earth's surface. Other systems are built on interactions between living things, such as predator-prey relationships.

Scientists recognize that systems can be either open or closed. An *open system* allows matter and energy to enter and exit. A *closed system* keeps matter and energy inside of it. Figure 1-1 illustrates both types of systems.

Open system

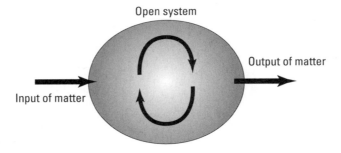

Input of matter

Output of matter

Closed system

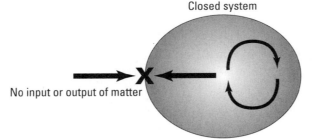

Figure 1-1:
Open and closed systems.

No input or output of matter

Illustration by Wiley, Composition Services Graphics

Very few systems in the natural world are truly closed systems. Scientists view the planet as a closed system in terms of matter (no matter enters or leaves the Earth), but they consider it an open system in terms of energy (energy enters the Earth from the sun). The following sections introduce you to a few of the Earth's other systems that you need to be familiar with. (Part II goes into a lot more detail on the different systems on Earth.)

Sorting the world into climate categories

One of the most important and complex systems that scientists study is the climate. The climate system includes but is actually much larger than local weather systems. Climate scientists observe how different parts of the Earth are warmed by the sun to greater or lesser degrees, and they track how heat from the sun moves around the globe in atmospheric and ocean currents.

The movement of heat and water around the Earth sets the scene for living things. Every living plant and animal has a preferred range of temperature and moisture conditions. The patterns of living communities on Earth are called *biomes*. Scientists define each biome according to its temperature and moisture levels and the types of plants and animals that have adapted to live within those limits. Understanding the complex link between climate factors and the distribution of life on Earth has become even more important as scientists document changes in the global climate and predict more dramatic changes to come. Turn to Chapter 7 for details on global climate patterns and biomes.

Dividing the Earth into ecosystems

Within every biome, scientists recognize various ecosystems, or communities of living organisms and the nonliving environment they inhabit. Studying how matter and energy move around ecosystems is at the core of environmental science. Specifically, scientists recognize that

- Matter is recycled within the ecosystem.
- Energy flows through an ecosystem.

Whether they're small or large, discrete or overlapping, ecosystems provide a handy unit of study for environmental scientists. Because plants are the energy base of most ecosystems (capturing energy from the sun), the type and number of plant species in an ecosystem determine the type and number of animals that the ecosystem can support. See Chapter 6 for details on ecosystems.

Observing the interactions between organisms within an ecosystem

Scientists called *ecologists* are particularly interested in how living things interact within an ecosystem. Plants and animals compete with one another for access to water, nutrients, and space to live. Evolution by natural selection has resulted in a wide array of survival strategies. Here are some examples (see Chapter 8 for more details):

- **Resource partitioning:** When two species, or types of animals, depend on the same resource, they may evolve behaviors that help them share the resource. This is called *resource partitioning*. An example is when one species hunts at night, while another hunts the same prey during the day.

- **Coevolution:** *Coevolution* occurs when a species evolves in response to its interaction with other species. Scientists have documented multiple cases of insects and the plants they feed on (and help pollinate) evolving to become more and more suited to one another over time.

- **Symbiosis:** Organisms that benefit from an interaction with another species live in what scientists call *symbiosis*. Symbiotic relationships between organisms may benefit both individuals, benefit only one while harming the other (such as with a parasite), or benefit one without harming the other.

Supplies Limited! Natural Resources and Resource Management

Environmental scientists do a lot of research to find ways to meet the needs of human beings for food, water, and energy. The environment provides these natural resources, but if their users (namely humans) don't care for them properly, they can be reduced, damaged, or destroyed. Managing natural resources for the use of human beings now while ensuring that the same resources will be available for humans in the future is called *conservation*.

Factoring in food, shelter, and more

People need food, water, air, and shelter to survive. But as human populations have grown into the billions, they've tested the ability of the environment to provide enough food, fresh water, and shelter. In Part III, I describe methods of sustainable agriculture and water conservation that can help meet the needs of so many people. (So far, there's still plenty of air to go around.)

Other resources that people depend on are less obvious, such as the biological diversity, or *biodiversity,* found in certain regions. Human actions have reduced biodiversity around the world, particularly in *biodiversity hotspots,* or regions with a combination of high levels of diversity and increasing human impacts. In Chapter 12, I explain what biodiversity is and why it's so important.

Thinking about energy alternatives

One of the most critical natural resources that modern living depends on is energy. Energy in most ecosystems streams from the sun every day, but to fuel modern life, humans have tapped into the stored energy of fossil fuels hidden deep in the Earth. Unfortunately, fossil fuel sources of energy are both limited in supply and damaging to the Earth's environment when humans burn them as fuel.

Searching for alternative sources of energy is an important part of environmental science research. Some of the current alternatives to fossil fuels include

- Solar energy
- Wind energy
- Hydro (river) energy
- Tidal and wave energy
- Geothermal heat
- Fuel cell electricity
- Liquid biofuel energy

I describe the pros and cons of these various options and explain how each one can help meet the energy needs of modern life in Chapter 14.

Keeping Things Habitable

Clean air, fresh water, food, and a safe place to live are critical to the survival of human beings. Unfortunately, in most parts of the world, decades of pollution have damaged environmental quality and endangered human health.

How humans can repair the damage already done to air, water, and land resources is the focus of Part IV.

Clearing the air (and water)

You may be familiar with some of the problems caused by air pollution: smog, acid rain, ozone depletion, and lung disease. In Chapter 15, I describe all the ways air is polluted and the results of pollution on ecosystems and human health. Similarly, in Chapter 16, I describe the sources and effects of water pollution.

In both cases, scientists classify the source of pollution as one of the following:

- ✔ **Point source pollution:** Point source pollution flows directly out of a pipe or smokestack and is easy to locate and regulate.

- ✔ **Nonpoint source pollution:** Nonpoint source pollution enters the air or water from a diluted or widespread area, such as when rainfall washes everything from city streets into nearby waterways via storm drains. This type of pollution is difficult to pinpoint and nearly impossible to regulate.

Tracking toxins and garbage

Toxic substances are all around you — in your home and in the environment. Many identified toxins today were once acceptable chemicals to use in agriculture or manufacturing. In some cases, scientists know the effects of a toxin, and as a result, it's no longer allowed to be used. In other cases, however, research is still being done to determine the danger of chemicals found in many household products.

In some places, toxins have entered the environment from improper waste disposal. Humans have to store (or burn) trash and other manmade garbage somewhere. All too often that garbage ends up in the oceans. I describe the problems related to waste disposal in Chapter 18.

When toxins enter an ecosystem, whether directly or as a byproduct of trash and hazardous waste, they can disrupt the ecosystem and cause harm to living things. Toxins often *bioaccumulate,* or build up in the cells of an organism. In some cases, the toxic substance is present in the environment at harmless levels but becomes more and more concentrated as it moves through the food chain. By the time top predators feed on lower predators, they've been poisoned by the *biomagnification* of the toxin. See Chapter 17 for more details on toxins and the effects they can have on the health of living things.

Influencing climate

These days, few environmental issues appear in the media and politics as often as modern climate change, or global warming. In Chapter 19, I explain how the greenhouse effect on Earth is beneficial and how greenhouse gases, both natural and manmade, change the composition of the atmosphere and affect climate patterns around the globe.

Some of the changes scientists expect with future climate warming include droughts in regions that are already water stressed, rising sea levels, and marine ecosystem disruption. The climate is definitely warming, so I also describe ways that humans can mitigate, or repair, the damage already done and adapt to a future climate that's very different from anything modern human civilization has experienced before.

Imagining the Future

Managing the Earth's resources so that human needs and desires today don't reduce the planet's ability to support future generations is called *sustainability*. The future is in your hands. The choices you make each day and the leaders you choose to create policies determine how people share, use, or abuse the Earth's resources in the coming decades. Regardless of your religious, political, cultural, or national values, you have a stake in your right and the rights of your children to a healthy, clean environment.

Realizing a sustainable economy

Many people think the biggest challenge in making sustainable choices is the cost, and some politicians want you to believe that a sustainable economy will destroy the world. Neither of these views is true. In Chapter 20, I describe some basic economic ideas and offer ways to look at the economy more sustainably. The transition to a more sustainable economy will take time, but in the long run, it'll be worth the effort!

Putting it on the books: Environmental policy

In Chapter 21, I introduce you to some of the most important and effective international agreements on global stewardship. The Montreal Protocol is one international agreement that was created to protect the environment. Specifically, this agreement reduced the production of ozone-damaging molecules around the world and halted the destruction of the ozone layer.

You may not have realized this, but 50 years ago many of the rivers, lakes, wetlands, and shorelines in the U.S. were much more polluted than they are today. After Congress amended the Clean Water Act in the 1970s, major cleanups began, improving water quality across the nation during the next few decades. These days new issues, such as climate change and environmental toxins, have taken a front seat in environmental science and policy. But no matter what issues are currently taking up the most attention on TV and in scientists' labs, the choices you and I make every day will determine the future health of the global environment.

Chapter 2

Lab Coats and Microscopes: Thinking Scientifically

....

In This Chapter

▶ Getting to know the scientific method

▶ Illustrating data with graphs

▶ Measuring the unknown

▶ Thinking critically about science in the media

....

*I*f you think of science as lab coats, microscopes, test tubes, pages and pages of data, and wild-haired scientists, you may be a little intimidated by the *science* part of environmental science. But in actuality, you perform acts of science every day; you just may not know it.

In this chapter, I describe what scientific thinking is, and I explain how scientists look at the world by asking and answering questions in an organized way (ahem, anyone heard of the scientific method?). I also introduce the most common ways that scientists and environmental scientists, in particular, present what they've learned by using graphs and statistics. Finally, I explain what good scientific news reporting looks like so you can evaluate the science you read about or see in the news.

Asking and Answering Questions with the Scientific Method

Scientists ask questions about the world around them just as you do. Is it cold outside today? Will a quarter of a tank of gas get me to work and back? Why do roses smell so good? Thinking scientifically simply means that when you ask a question, you go about answering that question in a methodical way, using logical reasoning. This way of asking and answering questions is often called the *scientific method*.

In this section, I describe the two approaches to logical reasoning, and I walk you through the various steps in the scientific method, including the ins and outs of designing experiments and the added step that professional scientists take — having their peers review their work.

Reasoning one way or another: Inductive versus deductive

Scientists construct their understanding of nature through logical reasoning, in which they follow a sequence of statements that are true to their conclusion. The two types of logical reasoning are

- ✔ **Inductive reasoning:** *Inductive reasoning* begins with a detailed truth about something and uses that truth to construct a generalized understanding of how the greater system or phenomenon functions. Using inductive reasoning to understand complex systems can be tricky because a few small details may not accurately represent the entirety of the system. Inductive reasoning is the opposite of deductive reasoning.

- ✔ **Deductive reasoning:** *Deductive reasoning* starts with broad generalizations and gradually focuses in on a specific statement of assumed truth. Deductive reasoning is most useful when you don't understand all the details of something but you can observe some of its outcomes. On the path of deductive reasoning, a scientist rules out one option after another until she has narrowed the field of truth down to just one or a few reasonable explanations. When a scientist proceeds with testing her hypothesis (see the next section), she's using deductive reasoning.

Working through the scientific method

Most students have encountered the scientific method at some point in their grade school education. For example, many teachers ask their students to write down each step they take while performing a lab experiment. In case you've forgotten a few things since grade school, I walk you through the main steps in the scientific method here:

1. **Make an observation.**

   An *observation* is just information you collect *empirically* (meaning that you collect the information by using your senses — sight, hearing, touch, taste, and smell) and *objectively* (meaning that anyone else in the same place, using the same methods, would observe the same information that you do).

Empirical and objective observations are what scientists call *data*. Scientists use data to create new hypotheses that they can then test by collecting more information, or *experimenting*.

2. **Create a hypothesis.**

   Based on your observation and any prior knowledge you may have from previous observations or experiences, you create a *hypothesis,* which is simply an inferred or assumed understanding based on your observations.

3. **Design and conduct an experiment.**

   After you have your hypothesis, you need to find a way to determine whether it's correct. Testing your hypothesis requires that you conduct an experiment (see the next section for details on this step).

4. **Analyze the results and draw a conclusion.**

   After you perform an experiment, you have more observations or data to incorporate into your overall understanding of your hypothesis. At this point, you may want to create a new hypothesis (if the data you collected during your experiment proved your original one to be wrong) and perform further experiments, or you may have enough new information to draw a conclusion, expanding your understanding of what you initially observed.

Although scientists use this method in their laboratories and in field settings where they collect scientific data on a daily basis, you use the scientific method every day without even realizing you're doing so. Take for example your morning shower: You turn on the water by adjusting the dial to what you think will be the right temperature and then you wait a few minutes:

- ✔ **Observation:** After a few minutes have passed, you observe steam forming around the flowing water.

- ✔ **Hypothesis:** You propose a hypothesis: The water is just the right temperature now.

- ✔ **Experiment:** Then you test the hypothesis with an experiment. You stick your hand in the water and observe the temperature.

- ✔ **Results and conclusion:** After you've collected data about the temperature of the water, you determine whether your hypothesis is true: Either the temperature is just right, or it isn't just right. If it's too hot, you infer that adding more cold water will make it the right temperature (and vice versa). Eventually, after you've collected enough data and made a number of inferences and adjustments, you'll find the water temperature that's just right for you to hop in the shower.

The point of this example is to illustrate that the scientific method isn't something magical or some kind of secret code. It's simply a way of describing how human beings ask questions and collect information to answer those questions. Environmental scientists use this methodical approach over and over again to understand the natural world.

Designing experiments

Experimental design is an extremely important part of the scientific method. When a scientist seeks to prove or disprove her hypothesis, she must carefully design her experiment so that it tests only one thing, or *variable.* If the scientist doesn't design the experiment carefully around that one variable, the results may be confusing.

The two main types of experiments scientists use to test their hypotheses are

- **Natural experiments:** *Natural experiments* are basically just observations of things that have already happened or that already exist. In these experiments, the scientist records what she observes without changing the various factors. This type of experiment is very common in environmental science when scientists collect information about an ecosystem or the environment.

- **Manipulative experiments:** Other experiments are *manipulative experiments,* in which a scientist controls some conditions and changes other conditions to test her hypothesis. Sometimes manipulative experiments can occur in nature, but they're easier to regulate when they occur in a laboratory setting.

  Most manipulative experiments have both a *control group* and a *manipulated group.* For example, if a scientist were testing for the danger of a certain chemical in mice, she would set up a control group of mice that weren't exposed to the chemical and a manipulated group of mice that were exposed to the chemical. By setting up both groups, the scientist can observe any changes that occur only in the manipulated group and be confident that those changes were the result of the chemical exposure.

  When designing manipulative experiments, scientists have to be careful to avoid *bias.* Bias occurs when a scientist has some preconceived ideas or preferences concerning what she's testing. These ideas may influence how she sets up the experiment, how she collects the data, and how she interprets the data. To avoid this bias, a scientist can set up a *blind experiment,* in which other scientists set up a control group and a manipulated group and don't inform the scientist who's actually observing the experiment which one is which.

Using scientific models to test hypotheses

You may have heard your local weather reporter talk about weather models and model predictions for the weekend ahead. Or maybe you've heard about climate models and their predictions of future climate change (see Chapter 19 for details). In both cases, the models may seem like crystal balls that can predict the future. However, climate and weather models are powerful tools that scientists use to understand complex global systems and predict how those systems will act in the future.

In some ways, a *scientific model* is very much like a model train or airplane in that it has parts that represent all the details of real life and some models are more detailed than others. Regardless of how detailed scientific models are, scientists can use them to test their hypotheses when studying the real thing is too difficult or, in some cases, impossible.

Take for example a globe: The continents, national borders, and locations of water, mountains, and other features represent what scientists know about the Earth but couldn't observe directly before satellite pictures were possible. A model of the Earth can be just one piece in a more complex model of the solar system, such as the one illustrated in Figure 2-1.

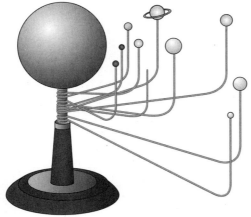

Figure 2-1:
A model of the solar system.

Illustration by Wiley, Composition Services Graphics

These days many scientific models are computer models (rather than physical models like globes) because computers can combine and analyze huge amounts of data much faster than a human brain. In environmental science, you're likely to encounter numerous detailed and complex models, such as climate models and ecosystem models. Scientists use these complex models to teach others (including you and me) about how large, intricate systems in the natural world function.

Before any scientific model (especially a complex computer model) can be informative, scientists have to do a lot of research to make sure it accurately models interactions in the real world. As part of their research, scientists set up the model and input data about the real world that led to a known event, such as a hurricane. If the computer model, using the real-world data, creates wind patterns, temperatures, wind speeds, and rainfall that are similar to or the same as the conditions in the real hurricane that occurred, then the scientists can be confident that the model will accurately simulate real life.

Asking peers for constructive criticism

One major difference between your everyday use of the scientific method and the way professional scientists use it is that you don't have to ask all your friends whether they agree that your shower water temperature is just right. Professional scientists, on the other hand, do have to ask for their peers' opinions about their experiments and the conclusions they draw from them. They do this through a process called *peer review*.

Any time a scientist completes a research project, she writes a paper or article about it, in which she describes her hypothesis, experiment, results, and conclusions. She sends this article to a peer-reviewed journal, where other scientists who are specialists in the same type of science review her work and determine whether the methods in the experiment make sense and whether the data she collected support the conclusions she made.

If the scientist's peers think she has done good work, then her article gets published in the journal so that the broader scientific community can learn from what she did. If her peers think her methods need improvement or her conclusions don't make sense, then they tell her that she needs to improve, clarify, or otherwise revise her study to be more accurate.

After a scientific study has passed the initial peer-review stage and has been published, it's open for debate among the broader community of scientists and can be very helpful to other scientists who are asking similar questions. In this way, scientists across the globe build a large database of information gathered from all of their experiments.

The careful approach to experimenting and collecting data that scientists have to take to be deemed credible, as well as the rigorous process of peer review they undergo after every major research project, results in scientific progress that appears to move at a snail's pace. This seemingly slow progress can be extremely frustrating for the general public and policymakers, who want to make decisions now and don't want to wait until scientists have studied everything scientifically from every angle. (In Chapter 20, I explain the different approaches some policymakers take in light of this.)

Speaking scientifically

Some words that you use in everyday conversation have a very different meaning when you use them to describe science. Here are a few words you may think you know the meaning of, along with what they mean when a scientist uses them:

✔ **Hypothesis:** A *hypothesis* is based on observations and states an assumed fact in a way that it can be tested. When scientists are working to rule out incorrect ideas, they may propose a *null hypothesis*. A null hypothesis usually says something like this: There is no relationship between how much I turn the hot water knob and the temperature of my shower. Testing this hypothesis will prove the null hypothesis false because there is, indeed, a relationship between how much you turn the hot water knob and the temperature of your shower. In some cases, ruling out wrong ideas is an important part of the deductive reasoning process.

✔ **Theory:** Scientifically speaking, nothing is ever "just a theory." A *theory* in science is an explanation of a natural phenomenon that has been tested repeatedly and is currently accepted as a fact. Theories explain why things occur the way they do or offer a *mechanism* behind what scientists observe. Theories (like any scientific findings) are open to further testing, but a good theory continues to hold true through these tests.

The scientific community isn't quick to call something a theory, so when it uses the term, such as for the theory of evolution by natural selection in biology, the theory of gravity in physics, or the theory of plate tectonics in Earth science, whatever phenomenon the theory is describing is considered an accepted scientific fact to be improved upon with further testing and observation.

✔ **Law:** A *scientific law* explains a process that has been observed, but it doesn't explain why the process occurs. For example, the law of gravity explains how two objects are attracted to one another, such as how the moon is attracted to the Earth, but it doesn't explain why the two objects are attracted to one another. To answer that question, you need to refer to the theory of gravity. *Note:* People often make the mistake of thinking that a scientific theory will someday become a law, thus making it more true or more factual. But that's not the case. Theories never become laws; they're two completely different things.

Presenting Data Graphically

One of the challenges that all scientists (especially environmental scientists) encounter is how to share the knowledge they gain from their research and experiments with other people in a way that's easy to understand. Reading page after page of detailed scientific findings can be difficult or boring even for a scientist. So scientists often turn to more visual ways — namely, graphs — to illustrate their discoveries, share their data, and explain what they've found out from their experiments. Scientists can choose from a number of different ways to present data as graphs depending on the type and amount of information they have. Here are three of the most common ones:

✔ **Pie charts:** *Pie charts* are circles with pieces that represent different categories of some type of information. Each slice of pie may be smaller or larger than the others, but all together the slices equal 100 percent. Figure 2-2 shows a pie chart that presents the different items that compose the municipal solid waste (trash) recycled in the U.S. in 2008.

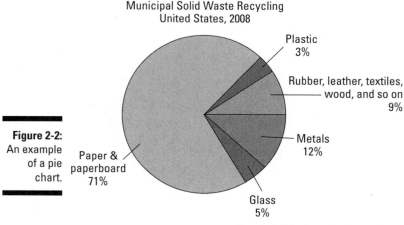

Municipal Solid Waste Recycling
United States, 2008

Plastic
3%

Rubber, leather, textiles,
wood, and so on
9%

Metals
12%

Glass
5%

Paper &
paperboard
71%

Figure 2-2:
An example
of a pie
chart.

Illustration by Wiley, Composition Services Graphics

✔ **Bar graphs:** A *bar graph* illustrates categories of data across a numbered scale. Bar graphs are also called *histograms* or *frequency distributions*. One axis of a bar graph represents the categories, and the other axis quantifies the different categories. Figure 2-3 shows a bar graph of the amount of trash produced by U.S. households for three different years. Each bar represents the total amount of trash for a particular year (the category of data), and the axis on the left provides a numerical scale to quantify each category. The bar graph in Figure 2-3 also presents the breakdown within each category of where the trash ends up: landfill, energy production, recycling, or compost.

✔ **Line graphs:** Scientists most often use *line graphs* to illustrate how measurements of a particular variable or data type change over time. The line graph illustrated in Figure 2-4 shows how the amount of recycling changed each decade from 1960 to 2006.

Pie charts, bar graphs, and line graphs may be the most commonly used graph types, but they certainly aren't the only ones. Regardless of the type of graph you're dealing with, the most important thing to do anytime you encounter a graph is to take a few minutes to read the title and axis labels. That way, you can more clearly understand what the author of the graph is trying to illustrate.

Municipal Solid Waste
per Household
United States, 1990–2008

Figure 2-3:
An example
of a bar
graph.

□ Waste landfilled ▨ Waste to energy ▩ Recycling ■ Composting

Illustration by Wiley, Composition Services Graphics

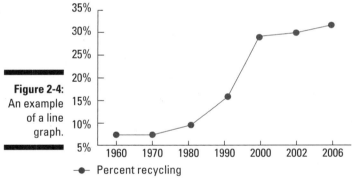

Percent Recycling Rates 1960–2006

Figure 2-4:
An example
of a line
graph.

—●— Percent recycling

Illustration by Wiley, Composition Services Graphics

Quantifying Uncertainty

Along with understanding the natural world, science also seeks to reduce the degree of uncertainty about what scientists know. Each scientific discovery through experimentation has some *degree of uncertainty,* meaning that some details are still unknown. By completing more studies and gathering more data, scientists gain more knowledge and reduce that uncertainty a little bit at a time.

Scientists use *statistics* (a mathematical way of interpreting data) to help them *quantify,* or measure, the amount of uncertainty involved. Certain statistical methods can determine whether the results they see are likely to be random (in which case the uncertainty is high and they still have a lot of unknown information to collect) or likely to be the result of the studied interaction (in which case the uncertainty is low).

One statistical tool to quantify the amount of uncertainty involved in a particular experiment is the probability value, or *p-value*. A p-value of 0.95, or 95 percent, means that scientists are 95 percent certain that the results of their experiment aren't random — in other words, that the results are very likely to be the result of an interaction between the variables. A p-value of 95 percent is an excellent result for most experiments.

Quantifying or measuring the uncertainty of scientific findings is important because before people use scientific knowledge to create policy or change the current way of doing things, they want to be as certain as possible that the data aren't the result of random chance.

Recognizing Good Science When You See It

In today's culture of fast-paced media, blog posts, blurbs, sound bites, and talking heads, knowing how to spot good science when you see it is more important than ever. The ability to distinguish between reliable information and unreliable information in the media is called *information literacy.* As a global citizen and a citizen of your community, you can use your information literacy to assess environmental issues and the scientific data used to make policy decisions.

To increase your information literacy, you need to be aware of the three types of sources for scientific information:

- **Primary sources:** *Primary sources* have the most recent and newly acquired scientific knowledge, and they've been evaluated by multiple other scientists to ensure that their methods are appropriate and their conclusions are logical. Peer-reviewed journals are an example of primary sources (see the earlier section "Asking peers for constructive criticism" for details).

- **Secondary sources:** *Secondary sources* explain the information from primary sources in a way that average readers can better understand it. Magazines, newspapers, and books are all examples of secondary sources. Articles in secondary sources aren't peer reviewed and can contain bias from the author or editor of the publication. They may also contain mistakes in how they interpret or represent the results of the scientific article.

✔ **Tertiary sources:** Even farther from the original source of information are *tertiary sources,* such as blogs and news commentary. These sources of information include a heavy dose of opinion by the writer or commentator. Tertiary sources can be a great place to learn the impact of new scientific knowledge on the cultural or political landscape, but often the scientific facts get lost in the heated debates and strong opinions. These are the least reliable sources of scientific information.

Table 2-1 summarizes these three types of information sources so you can more easily see how they're different.

Table 2-1	Sources of Scientific Information	
Source Type	*Characteristics*	*Examples*
Primary	Peer reviewed and includes technical details	Scientific journals
Secondary	Easier to understand for non-scientists and may have some errors in interpretation	Magazines, newspapers, books
Tertiary	Includes opinion, is likely to have more errors, and is far removed from the original source	Commentary, blogs, editorials

The best way to improve your information literacy and become more adept at recognizing bad science when you see it is to get out there and read. The next time you read a magazine, newspaper, or online article about any science-related issue, look for the following characteristics to know that your source is presenting you with good scientific information:

✔ The author cites the primary source (scientific journal) for any data presented.

✔ The author of the article identifies people quoted by name and professional association (where they work, who they work for, and similar details).

✔ At least some of the people quoted in the article are scientists or researchers in the field being discussed.

If your source doesn't include any of these characteristics, consider the information they present with a healthy dose of skepticism. Fortunately, these days so much knowledge is available on the Internet that if the topic really interests you, searching key words online may help you find primary or secondary sources that provide the scientific facts you're looking for.

Chapter 3

What's the Matter? Living and Nonliving Material

*B*efore you can fully understand the interaction between visible things, such as plants and animals in the environment, you need to understand the very smallest bits of matter — the ones you can't see — and the ways in which they interact with one another. After all, how these tiny bits of matter interact determines the structure and characteristics of the whole environment.

In this chapter, you find out how fundamental interactions between atoms and molecules lay the foundation for everything you see around you. Specifically, you find out which characteristics distinguish living matter from nonliving matter in the environment.

Changing States of Matter

Matter, which makes up everything around you, exists in three basic forms or *states:* solid, liquid, and gas. Take water for example. At a wide range of temperatures on the Earth's surface, water can exist as a solid (ice), a liquid (water), or a gas (steam). In each of these forms, the water itself contains the same matter; the matter just has different physical characteristics.

Like water, other matter changes states when you heat it or cool it, but the temperatures at which these changes take place vary depending on the matter. For example, alcohol evaporates (turns into a gas) at normal room temperature and must be much, much colder than water in order to freeze (about 100 degrees colder). Rocks, on the other hand, are solid at the range of temperatures on Earth's surface, but they melt into a liquid state deep beneath the surface, where temperatures are much hotter.

Matter changes states, but it never disappears or is created. This idea is called the *law of conservation of matter,* and it applies in particular to a closed system like Earth's environment (I describe closed systems in Chapter 6). In other words, every atom of matter that exists now on Earth has always existed and will always exist.

Examining Atomic Structure

All the matter in the universe is composed of small particles called atoms. An *atom* is the smallest bit of matter that represents one of the elements on the periodic table of elements. This table, illustrated in Figure 3-1, organizes all the known, naturally occurring (and some manmade) matter according to what characteristics its atoms have and how its atoms behave when reacting with other atoms.

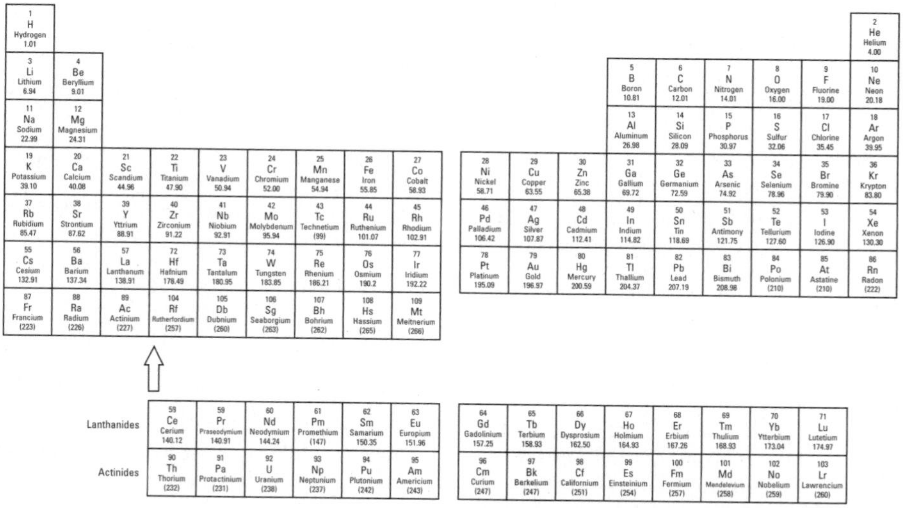

Figure 3-1:
The periodic table of elements.

Illustration by Wiley, Composition Services Graphics

Each atom is composed of *subatomic particles*. The *nucleus* (or center) of an atom is composed of subatomic particles called *protons* and *neutrons,* which are surrounded by at least one *electron shell,* where other subatomic particles called *electrons* are located. Figure 3-2 illustrates the parts of an atom.

Protons have a positive or +1 electrical charge, neutrons have a neutral or zero charge, and electrons have a negative or –1 charge. Every atom has an equal number of protons and electrons so that the atom itself has a neutral charge (the +1 and –1 cancel each other out).

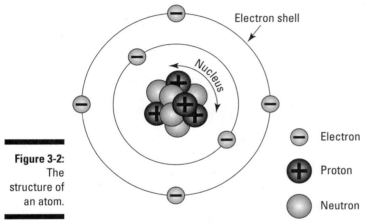

Electron shell

Nucleus

Electron

Proton

Neutron

Figure 3-2:
The structure of an atom.

Illustration by Wiley, Composition Services Graphics

During some chemical interactions, an atom may lose or gain electrons and become an atom with a plus or minus electrical charge. An atom with an electrical charge is called an *ion.*

The number of electrons an atom has in its outer electron shell determines how the atom forms bonds with other atoms, a process that I describe in the next section.

Making and Breaking Chemical Bonds

In order for atoms to construct matter large enough for you to see or feel, they have to combine with other atoms to form *molecules.* In fact, very few atoms exist all by themselves; they usually link together to create larger molecules. Sometimes atoms simply pair up with other atoms of the same element (as is the case in O_2, or oxygen gas), and other times they combine with atoms of other elements to form *compounds* (as is the case in H_2O, or water, which is made up of two hydrogen atoms and one oxygen atom).

To form molecules, atoms must exchange or share electrons from their outer electron shell to create *atomic bonds*. In the following sections, I describe the most common atomic bonds that you need to be familiar with when studying environmental science. As a bonus, I discuss a few ways atoms interact with one another without actually bonding, through the processes of oxidation and reduction.

Ionic bonding

An *ionic bond* occurs when one atom gives an electron to another atom. Atoms linked together in this way are called *ionic compounds*. The ionic compound you're most familiar with is table salt, which forms as the result of an ionic bond between the elements sodium (Na) and chlorine (Cl). When a sodium atom and a chlorine atom bond ionically, an electron from the sodium atom jumps over to the chlorine atom. When the electron exchange takes place, the sodium atom becomes a positively charged ion, and the chlorine atom becomes a negatively charged ion. Just like magnets with opposite electrical charges, the sodium and chlorine atoms cling together. Figure 3-3 shows you what this ionic bond looks like.

Understanding the basics of ionic bonding is important in environmental science because, although ionic bonds are strong, they're easily broken in water. I explain why in the later section "Surveying the Properties of Water."

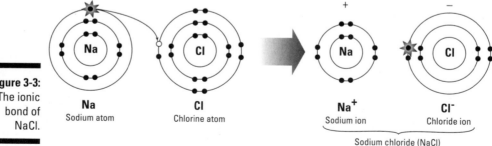

Figure 3-3:
The ionic bond of NaCl.

Na
Sodium atom

Cl
Chlorine atom

Na^+
Sodium ion

Cl^-
Chloride ion

Sodium chloride (NaCl)

Illustration by Wiley, Composition Services Graphics

Covalent bonding

When two atoms join together in a *covalent bond,* they form a molecule that shares electrons. Unlike in the ionic bond, neither of the atoms in a covalent bond loses or gains an electron; instead, both atoms use a pair of shared electrons.

The simplest covalent bonds form between atoms of the same element. For example, two oxygen atoms join together in a covalent bond to form a molecule of O_2 gas (see Figure 3-4 for a visual of this molecule).

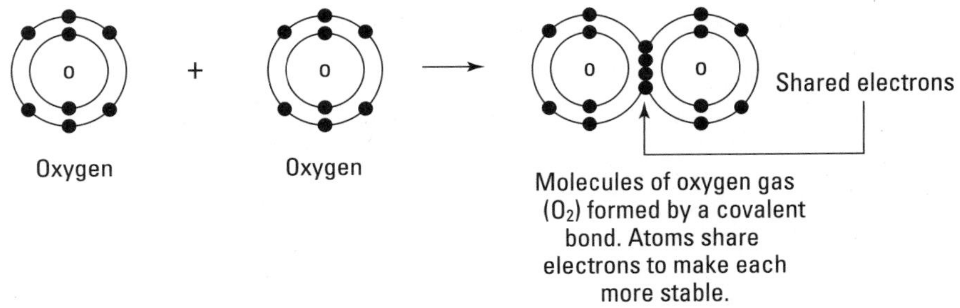

Figure 3-4: Covalent bond between two oxygen atoms.

Oxygen + Oxygen →

Shared electrons

Molecules of oxygen gas (O_2) formed by a covalent bond. Atoms share electrons to make each more stable.

Illustration by Wiley, Composition Services Graphics

To help predict which atoms are likely to form covalent bonds, scientists use the *octet rule.* This rule states that atoms are most stable (and, thus, nonreactive) when their first electron shell has two electrons and any outer shells have eight electrons.

Atoms that form covalent bonds are slightly unstable because they don't have enough electrons in their outer electron shell. But when one slightly unstable atom combines with another slightly unstable atom and they share their unstable electrons with each other, they both become more stable. As a result, the molecule the two atoms created is unlikely to react with other atoms around it; in other words, it becomes less *chemically reactive.*

Understanding covalent bonding is particularly important when you study *organic matter,* or matter that contains carbon (which I discuss later in this chapter). The element carbon has four electrons in its outer shell and can form covalent bonds with up to four other atoms at one time. By forming so many covalent bonds, carbon molecules can build large, complex shapes.

Hydrogen bonding

Another type of atomic bond you need to be familiar with when studying environmental science is the hydrogen bond. A *hydrogen bond* results when some of the atoms in a covalently bonded molecule pull the shared electrons to one side of the molecule, creating an electrical imbalance in the molecule. (Remember that electrons have a negative electric charge.)

When an imbalance of electrical charge occurs within a molecule, the molecule is said to be *polar* or to exhibit *polarity* — in which case one end has a positive charge and the other has a negative charge (just like a magnet). A polar molecule acts a little like an atom in need of an electron. Its positively charged end is attracted to negatively charged things around it — most commonly the negatively charged sides of other polar molecules.

The most common example of hydrogen bonding involves water molecules. Figure 3-5 illustrates how the covalent bond between oxygen and hydrogen creates an electrical imbalance in water molecules and how, as a result, the negative end of one water molecule is attracted to the positive ends of another water molecule.

Hydrogen bonds are weak compared to the covalent and ionic bonds I describe earlier, but these weak bonds play an important role in the environment and in living things. They're important in forming DNA chains, and they give liquid water some seemingly magical properties, which I describe in the later section "Surveying the Properties of Water."

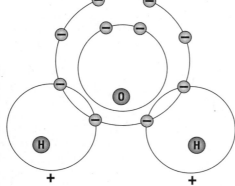

Figure 3-5:
Water
molecules
with slight
polarity form
hydrogen
bonds.

Illustration by Wiley, Composition Services Graphics

Oxidizing and reducing compounds

Many chemical reactions in the environment occur because an atom takes electrons from another nearby atom. Scientists call this type of reaction *oxidation* because oxygen was the first element recognized to react this way with other atoms and molecules. In an oxidation reaction, the atom that loses the electrons is said to be *oxidized*.

For example, rusted iron is the result of an oxidation reaction between oxygen (O) and iron (Fe) that produces the red-colored compound iron oxide (Fe_2O_3). Another example is when a carbon atom is oxidized as it combines with oxygen, forming carbon dioxide (CO_2).

Of course, sometimes the reverse reaction occurs, and an atom captures an electron from another one. The reverse of oxidation is called *reduction*. In a reduction reaction, the atom that receives the electron is said to be *reduced*.

One of the most important oxidation-reduction reactions occurs in the cycles of photosynthesis and respiration, which I describe in detail in Chapter 4. When plants perform photosynthesis, they reduce carbon dioxide, releasing oxygen into the atmosphere. When animals breathe in oxygen through respiration, they use it to oxidize the carbon compounds they've eaten in food and release carbon dioxide.

Surveying the Properties of Water

As a small molecule made up of three covalently bonded atoms, water exhibits a few unique characteristics that quite simply make life possible. In this section, I explain how the details of molecular bonding within and between water molecules create conditions that help living things on Earth thrive.

Taking a closer look at the water molecule

You may take water for granted as you drink it, wash it down a drain, and carry a large amount of it around in your body, but water is something special. Similarly structured molecules (such as hydrogen sulfide, or H_2S) exist only in gas form on Earth's surface. But water can exist as a solid, liquid, or gas in the range of temperatures on Earth's surface; this fact is what makes life on Earth possible.

As I describe earlier in this chapter, water molecules commonly take part in hydrogen bonding as a result of being slightly polar. Although the hydrogen bonds between water molecules are weaker than covalent or ionic bonds, they help water do things like break apart other molecules and cling to one another or to other slightly polar materials. The following sections describe a few ways the polarity of water molecules shapes how water behaves in the environment.

Acting as a solvent

A *solvent* is a liquid that dissolves or breaks down a solid when it touches that solid. To be a solvent, the molecules in water create bonds with the solid material that are stronger than the bonds holding the solid material together; as a result, the water molecules break apart the solid.

Because of its polarity, water is an excellent solvent for anything that's held together with ionic bonds. Just think about the many times you've dissolved salt in water. When you add the solid salt compound to water, the polar water molecules attract the ions of sodium (Na^+) and chlorine (Cl^-) and split the salt ($NaCl$) molecule apart, or dissolve it.

This characteristic of water is very useful to you when you're cooking or sweetening your tea. However, the ability of water to act as a solvent also means that dangerous substances can dissolve in water — polluting it or making it poisonous — and then circulate throughout the environment.

Creating surface tension

Water molecules cling together as a result of hydrogen bonding due to their polarity. This creates *surface tension,* which is a layer of bonded molecules that form a barrier or sort of film around the edge of the water.

You may have noticed that a drop of water on a tabletop often appears as a half-dome shape. This shape occurs because the water molecules cling together and create surface tension around them. The same property allows some insects to walk across the surface of water.

Moving upward with capillary action

Water exhibits a property called *capillary action,* in which its polarity leads it to be attracted to things around it that are also slightly polar, or charged. When you dip the corner of a paper towel in water, capillary action is what draws the water up into the towel. Similarly, when you walk outside on a rainy day, the hem of your jeans soaks water a few inches up your pants legs even though only the very bottom edge of your jeans was touching the wet street — capillary action again! The attraction of the water molecules to the new surface is so strong that it overcomes the pull of gravity, allowing the water to move upward into the paper towel or up the bottom of your pants legs.

Capillary action is what allows trees to take up water through their roots and circulate it all the way up to the leaves at the very top. It also plays an important role in moving groundwater through small spaces in rock and soil. (Find out more about groundwater in Chapter 9.)

The unbearable lightness of ice

The relationship between an amount of matter and the space it occupies is called its *density,* and it's usually measured in grams per cubic centimeter (g/cm^3). In most cases, when matter changes from a liquid to a solid, the same amount of matter takes up less space in solid form and, thus, becomes denser in solid form. In other words, the same number of molecules fill less space in solid form than they do in liquid form.

Water, however, doesn't behave the way other types of matter do in this regard. When water changes from liquid to solid, it takes up more space, thereby becoming less dense. This property results from the orderly arrangement of water molecules as they stack together, leaving space between them, as shown in Figure 3-6.

This property explains why an ice cube floats in your glass of water and why icebergs float in the ocean; the material with lower density (ice) floats in the material with more density (water).

Figure 3-6:
Water
molecules
in a solid
state (ice).

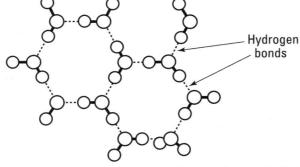

Hydrogen bonds

Illustration by Wiley, Composition Services Graphics

What's pH Got to Do with It? Acids and Bases

Many of the interactions that environmental scientists study are the result of *acid* and *base* reactions. So in this section, I provide a brief overview of the aspects of acid-base chemistry that you need to know as you study environmental science.

Defining acids and bases

So what are acids and bases, anyway? Empirically speaking, acids and bases have the following characteristics:

- ✔ Acids usually have a sour taste (like lemon juice) and will burn your skin.
- ✔ Bases usually have a bitter taste and feel slippery (like detergent).

Chemically speaking, an *acid* is a compound that gives away or releases a hydrogen ion (H^+) during a chemical reaction, and a *base* is a compound that captures a hydrogen ion during a chemical reaction.

The slight polarity of a water molecule, which I explain in the previous section makes it a powerful agent for dissolving acid and base compounds into liquid form, or *solution*. When water dissolves an acid compound, the compound releases positively charged ions of hydrogen (H^+) into the solution.

The number of available hydrogen ions in the solution determines how it will chemically react with other compounds. Scientists quantify the concentration of hydrogen ions in a solution by using the pH scale.

Measuring reactivity with the pH scale

The *pH scale* measures how acidic a solution is; it ranges from 1 to 14. Lower numbers on the pH scale indicate higher levels of acidity (more available hydrogen ions), while higher numbers indicate more basic solutions. A pH value of 7 is neutral, meaning that the solution is neither acidic nor basic.

The liquids you encounter everyday range from pretty acidic (lemon juice) to pretty basic (oven cleaner) and everything in between. Table 3-1 shows the approximate pH values of some common substances.

Table 3-1	Approximate pH of Common Substances
pH Value	*Substance*
14	Drain cleaner
13	Bleach
12	Soapy water
11	Ammonia
10	Antacid
9	Baking soda
8	Salt water
7	Water
6	Urine
5	Coffee
4	Soda, beer
3	Aspirin
2	Lemon juice
1	Battery acid

Environmental scientists need to understand pH because the acidity of a solution determines how that solution will react with the other matter it contacts. For example, when I discuss air pollution and acid rain in Chapter 15, I describe how it erodes and destroys limestone statues. This erosion is a result of the acid in the rain interacting with the limestone rock and dissolving part of it, ruining the statue in the process. In Chapter 19, I describe how increasing levels of carbon dioxide in the atmosphere lead to a lower pH of the oceans and how that lower pH affects marine life.

It's Alive! Organic Matter and Compounds

Biological molecules (or *organic matter*) make up a major component of the environment. Constructed of atoms and molecules, organic matter follows the same rules of chemical bonding as the inorganic matter I describe earlier in this chapter. However, what separates organic matter from inorganic matter is its ability to construct complex, living things, like you and me.

Organic compounds, sometimes called *carbon compounds,* are the basis for all living matter on Earth. Organic compounds are molecules made of atoms of carbon and a few other elements. All organic compounds share the following basic characteristics:

- ✔ The molecules are based on the unique ability of carbon atoms to create covalent bonds with multiple other atoms at the same time. (The complexity of carbon bonding is so vast that an entire scientific subdiscipline of chemistry, called *organic chemistry,* is devoted to studying it.)

- ✔ The molecules are primarily composed of hydrogen, oxygen, carbon, and nitrogen with very small amounts (less than 2 percent) of elements like calcium, phosphorus, and sulfur.

- ✔ The molecules are *modular,* meaning that a few different shapes can combine in multiple ways to create a wide variety of unique and complex results.

- ✔ The *geometry,* or three-dimensional shape, of the basic molecules determines how they combine to form larger, more complex compounds.

Scientists classify molecules that fit these descriptions into four main groups of organic compounds. These four groups, which I describe in the following sections, make up all living matter and control all the chemical processes of organisms.

Proteins

Proteins are made up of smaller organic compounds called *amino acids* that are linked together in chains. Many different kinds of amino acids can be built with carbon atoms, but living organisms produce only 20 different amino acids. From these 20, all the proteins needed for life are constructed.

Proteins play a variety of important roles in living organisms:

- ✔ **Proteins act as enzymes.** *Enzymes* assist other molecules during chemical reactions by providing structure. Sometimes enzymes are molecular matchmakers because they bring two (or more) molecules together and help them connect to form something more complex. Other times they help break molecules apart.

- ✔ **Proteins form hormones.** *Hormones* function as messengers between cells in an organism, signaling to different cells when to begin or end particular activities.

- ✔ **Proteins build antibodies.** *Antibodies* are what your immune system (and the immune system of every living thing) uses to fight off illness and disease.

   ✔ **Proteins provide structural support.** Proteins make up the structural components of organic matter (see the later section "Building Organisms One Cell at a Time" for details).

Nucleic acids

Nucleic acids are the organic molecules that contain instructions for cell reproduction and energy use. The most familiar nucleic acid is DNA, or deoxyribonucleic acid. DNA molecules link together to form genes, which are the blueprints for creating new cells with specific physical and functional characteristics.

In Chapter 12, I explain how biological change occurs through the combination of DNA from two parents to create their offspring.

Carbohydrates

Carbohydrates are molecules that help organisms store and use energy, as well as provide structural material. The simplest carbohydrates are single sugar molecules, such as *glucose.* Two examples of more complex carbohydrates are *cellulose,* which forms structures such as plant fibers, and *starch,* which plants use to store energy. When an animal eats a plant that contains starch, the animal breaks down the starch into its component glucose molecules to use as energy. (I offer plenty of details on energy in Chapter 4.)

Lipids

Lipids are organic compounds that don't dissolve in water; they include many greasy or oily substances, such as wax, butter, and oil. Lipids play two vital roles in living organisms:

   ✔ **To form watertight membranes:** The fact that lipids don't dissolve in water makes them very important in the structure of cell membranes (which I describe in the next section). Every cell needs a boundary to keep some things in and other things out. Lipids provide that boundary.

   ✔ **To store energy:** Although carbohydrate and protein molecules also store energy for organisms, neither of them does so as effectively as a lipid with its long chain of molecules that resists dissolving in water.

Building Organisms One Cell at a Time

Just as an atom is the fundamental unit of matter, a *cell* is the fundamental unit of life. A cell performs all the functions necessary to maintain life and create new cells, including capturing, using, and storing energy.

In the following sections, I describe living things at their most basic level, including organisms that consist of only one cell and others that consist of many different cells. I also explain the subtle but important differences between plant and animal cells and illustrate their internal components.

Working together: Cell specialization

The simplest organisms on Earth are single cells that function as a complete life form (using energy, growing, and reproducing) all by themselves. These single-celled organisms have been around for a long time and continue to thrive even though you can't see most of them without a microscope.

Most living organisms, and certainly most of the visible ones, are composed of many cells working together. The trick to these *multicellular organisms* is cell specialization.

Cell specialization is when cells take on different jobs or assignments within the organism. For example, your hair cells perform different functions than your skin or liver or eye cells. Similarly, in a plant, some cells form leaves and others form the stem or trunk.

Separating plant cells from animal cells

Regardless of which specialized job a cell plays in the larger organism, all cells are fundamentally similar in structure. In fact, the cells of both plants and animals have the same basic internal components. Here's a list of the parts, or *organelles,* common to every cell:

- ✔ **Cell membrane:** The cell membrane, which is built of lipids, defines the outer boundary of the cell. Within the cell membrane, other organelles are protected. Cell membranes prevent some materials from passing through, but they allow other materials (such as energy and waste) to pass into and out of the cell.

- ✔ **Nucleus:** The *nucleus* is where the genetic information (DNA) of the cell is stored.

- ✔ **Mitochondria:** The *mitochondria* in a cell provide fuel for everything the cell does. I explain this process, called *cellular respiration,* in Chapter 4.

In addition to these organelles, plant cells have a few other important parts that animal cells don't have:

- **Cell wall:** Along with the cell membrane, plant cells have a *cell wall* made of cellulose that helps the cell maintain a more rigid structure.

- **Chloroplasts:** *Chloroplasts* are organelles inside plant cells that use a molecule called *chlorophyll* to capture sunlight and transform it into energy. (I explain this process, called *photosynthesis,* in Chapter 4.)

Figure 3-7 illustrates a plant cell and an animal cell and includes the major components of each.

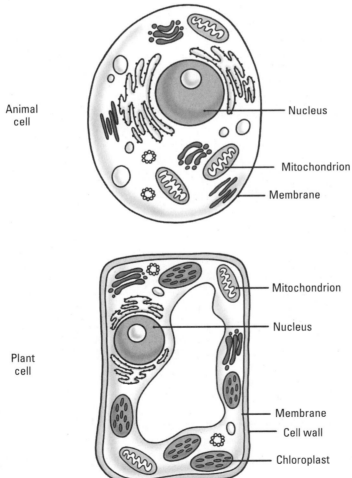

Animal cell — Nucleus

— Mitochondrion

— Membrane

Plant cell — Mitochondrion

— Nucleus

— Membrane

— Cell wall

— Chloroplast

Figure 3-7:
A plant cell and an animal cell.

Illustration by Lisa Reed

REMEMBER The most significant difference between plant and animal cells is that animal cells don't have chloroplasts, which means they can't capture and convert energy from the sun the way plant cells do. Instead, animal cells must get their energy from outside of themselves.

Living an extreme life

For hundreds of years, scientists classified living things into one of two kingdoms: animal or plant. But as scientists recognized more diversity among living things, they added three more kingdoms to include *bacteria* (single-celled organisms with simple cells), *fungi* (molds, mushrooms, and so on), and *protists* (single-celled organisms with complex cells). Then in the 1970s, this five-kingdom classification system was turned on its head when researchers discovered a new life form — the archaea.

Archaea are single-celled organisms usually found in extreme environments on Earth that scientists previously thought couldn't support life. These single-celled organisms are similar to bacteria in appearance, but genetic studies indicate that they're as different from bacteria as they are from plants and animals. As a result, scientists now use a three-domain classification scheme for living things: bacteria, archaea, and *eukarya* (complex-celled organisms, including plants, animals, fungi, and protists).

Generally called *extremophiles* because they thrive in extreme environments, scientists label archaea according to the specific type of extreme environment they live in. Here are a few examples:

✔ *Hyperthermophiles* live in places with high temperatures, up to and exceeding the boiling point of water, including hot springs and seafloor vents.

✔ *Acidophiles* live in places with very low pH, or high acidity.

✔ *Alkaliphiles* live in places with very high pH, or extremely low acidity.

✔ *Cryophiles* live at temperatures below freezing, such as in glacier ice, in permafrost, and below winter snowpack.

✔ *Xerophiles* live in extremely dry conditions, such as in deserts.

✔ *Halophiles* live in environments with high concentrations of dissolved salt.

Scientists have found other archaea in environments with high sugar concentrations and high pressure conditions and even inside rocks and oil below the Earth's surface. In fact, as they continue their research, scientists are finding that archaea are everywhere on Earth, inhabiting environments from one extreme to the other and everything in between.

One of the most interesting things archaea have taught scientists is that life can exist under some pretty strange conditions. For scientists who look for organisms in other parts of the solar system (called *astrobiologists*), the discovery of archaea on Earth opens up new possibilities of what may survive on other planets.

Chapter 4

Making Things Happen: Energy

*E*nergy is at the heart of many issues in modern environmental science, from finding sustainable energy sources to fuel things like modern technology (cars, televisions, cellphones, and so on) to maintaining a healthy energy flow through ecosystems. (Turn to Part II for more on ecosystems.)

Because energy is such an important part of environmental science, I dedicate this whole chapter to helping you understand what energy is, what it's used for, and where it comes from. I describe the scientific understanding of energy, starting with scientific laws and moving into a quick overview of how it's transferred from one place to another or from one organism to another. I finish up with a quick look at photosynthesis, the process that turns the sun's energy into fuel for life on Earth's surface, and respiration, the process that releases that fuel for use.

Identifying the Different Types of Energy

Energy may seem like a vague concept — something magical that makes things happen. In a sense, it is a little magical; after all, even physicists admit that it's difficult to explain. But when you're studying the environment, you can think about *energy* as the ability of matter to move from one place to another or from one state to another. As it flows from one object to another, energy changes the behavior of those objects, making them warmer or setting them in motion. Specifically, you can examine energy as either heat or work:

✔ **Heat:** The flow of thermal energy between two objects of different temperatures

✔ **Work:** The flow of energy that physically moves an object

In this section, I describe specific ways that energy flows from one object (or material) to another and the different forms of energy scientists have identified.

Transferring heat: Thermal energy

Heat, or *thermal energy,* isn't the same as temperature; rather, heat is the flow of energy between two objects of different temperatures. For example, when you touch something that feels hot, you're experiencing a transfer of thermal energy (heat) due to the difference in temperature between your hand and the object.

The random motion of atoms and molecules moving around and bumping into one another creates thermal energy. Although the motion of the atoms is much too small for you to see, you can feel it as heat. The more atomic activity an object has — that is, the faster and more furiously its molecules are moving — the hotter the object feels. The thermal energy of molecules in an object affects everything the object touches through *heat transfer* — the process of increasing the activity at the atomic level of other materials or objects.

Heat moves between objects in three ways, which I describe in the following sections. These three ways — conduction, convection, and radiation — are all illustrated in Figure 4-1.

Figure 4-1: Movement of heat via conduction, convection, and radiation.

Water at base of pot is heated by conduction and then moves upward, creating convection cells, which carry heat to the surface of the water.

Radiation of heat in waves.

Conduction of heat energy from stove burner to pot.

Illustration by Lisa Reed

Conduction

Conduction occurs when a heated material physically touches another material, thus transferring heat (thermal energy) to that material. How does conduction work? The process is quite simple. The highly mobile atoms in the hotter material bump into the atoms of the cooler material, causing them to start moving faster and, thus, heating them up.

Figure 4-1 shows an example of heat transfer by conduction, where a stove burner is touching the bottom of a pot of water. The stove burner is conducting thermal energy to the pot, heating it up.

Conduction occurs throughout the environment anytime two objects at different temperatures come into contact with each other.

Convection

Convection is the process of transferring heat (thermal energy) through the motion of a material (most often a liquid or gas). Convection occurs when matter is heated and the hotter atoms in that matter move faster, spreading out and filling more space. When matter expands in this way, filling up a larger space, it becomes less dense. This heated, less dense material moves upward in relation to the cooler, denser material that surrounds it.

The motion of the hotter material moving upward and the cooler material moving downward creates a circular motion, called a *convection cell*. Figure 4-1 shows a convection cell in a pot on a stove top. After the stove heats the bottom of the pot, the bottom of the pot heats the water in the pot (through conduction), making the water less dense. The heated water then moves upward, forcing the cooler water near the top downward. When the hot water reaches the top of the pot, it cools, becoming denser again, and the newly heated, less dense water from the bottom of the pot moves upward, displacing the newly cooled water. In this way, through the density-driven motion of heated material, heat is transferred from the bottom of the pot to the top by the convective motion of the water.

How does convection occur in the environment? Convection cells transfer thermal energy throughout the oceans and atmosphere, creating the weather you experience every day. See Chapter 7 for details on climate.

Radiation

Heat energy can also flow as a wave. Heat transferred as a wave is called *radiant heat*. Unlike conduction, radiant heat transfer (or *radiation*) doesn't require atoms and molecules to be physically touching to transfer or move heat. And unlike convection, radiation doesn't require a change in the density of the matter as it transfers heat.

You experience radiant heat when you hold your hands near (but not touching) a hot pot on a stove burner. You feel heat radiating from the pot. Some of the heat you feel is the conduction of heat from the stove to the air surrounding it, but most of the heat comes from the radiant waves of heat moving out from the pot. Figure 4-1 shows radiant thermal energy being transferred outward from the hot pot. (You also experience radiant heat when you hold your hands near the stove burner itself. Notice that the conduction shown in the figure comes from the pot actually touching the burner; you'd experience radiant heat if you held your hand a little away from the burner.)

While heat transfers change the temperature of matter, a different kind of energy is needed to really get things moving. Keep reading to find out how work sets things in motion.

Setting things in motion: Work and kinetic energy

The energy of motion is called *kinetic energy.* To increase or decrease the kinetic energy of an object, you apply a *force,* such as a push or pull. When a force is applied, energy (called *work*) transfers into the object and increases its kinetic energy, or motion. Scientists define work as the force required to move an object over a distance.

Often, the motion (kinetic energy) of one object directly transfers motion (kinetic energy) to another object through physical contact, such as when billiard balls hit one another.

Storing potential energy

When energy isn't flowing between objects, heating them, or setting them in motion, it can be stored for later use as *potential energy.* To understand how kinetic energy transfers into potential energy, think about a rubber band. When you hold a rubber band with two hands and pull back on one end, the motion (kinetic energy) of your pulling transfers energy to the rubber band, and that energy is then stored as potential energy in the stretched rubber band. After you release your hold on one end of the rubber band, the potential energy is released as kinetic energy, transporting the rubber band through the air. In this example, the stored energy is called *elastic potential energy* because the stored energy is held in the stretch of the rubber band. (Elastic potential energy plays an important role in transferring kinetic energy in waves, which I describe in the following section.)

In environmental science, you're more likely to encounter potential energy stored in chemical bonds or *chemical potential energy*. To release chemical potential energy, you have to break the atomic bonds that hold chemical compounds together (see Chapter 3 for details on atoms, molecules, and bonding). Consider this example: In an automobile, chemical potential energy is stored in the gasoline. As the gasoline breaks down, it releases its potential energy as kinetic energy, which moves the parts of the car engine, which, in turn, moves the car.

You also encounter chemical potential energy when you eat food because the food contains stored energy in its chemical bonds. Your body then turns the chemical potential energy into kinetic energy to fuel your daily life. I explain more about this process in the section "Converting the Sun's Energy to a Useable Form" later in this chapter.

Moving in waves

Waves of kinetic energy, called *mechanical energy waves,* can only move through a *medium* (some kind of matter). These mechanical waves transfer energy across a distance through matter without carrying the matter along. As mechanical waves move through matter, they shift the atoms and molecules up and down (or side to side) in a wave motion as the energy flow alternates between elastic potential energy and kinetic energy through the matter.

To visualize how mechanical energy waves work, picture a ship adrift on stormy seas. While the waves roll by, the ship moves up and down with each wave, but none of the waves carry the ship too far. The waves move through the water, causing the ship to swell up and down in motion, but the water itself isn't being carried forward, so neither is the ship.

Here are two other types of mechanical energy waves that transfer kinetic motion:

- **Earthquake waves:** These waves move through the rock of the Earth's crust and interior, transferring the energy of an earthquake all around the globe.
- **Sound waves:** What you hear as sound is the vibration of air molecules as waves of energy move through them. Sound waves move through air, water, and other matter but can't move through the emptiness of space.

Wave energy that doesn't rely on matter to be transferred is called *radiation,* or *radiant energy,* and can move through the emptiness of space. Heat is one type of radiant energy, but radiant energy also exists as visible light, UV rays, X-rays, and other forms of wave energy along the *electromagnetic spectrum.*

The electromagnetic spectrum is the range in size of different types of radiant energy waves. The distance from the top of one wave to the next determines whether the wave is visible (like sunlight) or invisible (like X-rays or microwaves). Energy transfer through electromagnetic radiation is a result of a complex interaction between electricity and magnetism at an atomic level that creates waves.

Electromagnetic waves of visible light that flow from the sun provide the energy source for almost all living things on Earth's surface. Check out the section "Converting the Sun's Energy to a Useable Form" to find out how.

Defining the Thermodynamic Laws

Energy behaves in predictable ways, conforming to *laws of thermodynamics*. (The term *thermodynamics* simply means heat and motion.) Understanding the basic ideas behind these laws is helpful when you begin to examine environmental systems and energy flow through the environment (see Chapter 6 for details on both of these topics):

- **First law of thermodynamics (the law of conservation of energy):** This scientific law states that energy must come from somewhere. In other words, energy must be added to a system as heat or work in order to create changes (in temperature or motion) within the system. You may hear this law summarized as the idea that energy can't be created or destroyed and is, therefore, conserved (hence the name, *law of conservation*).

- **Second law of thermodynamics (the law of entropy):** This scientific law states that energy flows in one direction (for example, the way heat flows from hot materials to cold materials) but never the other way around. This continuous flow of energy in one direction leads a system toward randomness or disorder. As a result, systems need increasing energy added from outside the system to contain and control things and keep the systems ordered as the energy flows through them. The word *entropy* in this law simply means the tendency of something to proceed in one direction.

Counting Calories (And Joules)

Unless you're a physicist, your main interest in energy is most likely related to how much it costs and how much of it you use in your daily life. To determine how much energy costs or how much you use, you first need to understand how energy is measured.

Scientists define *energy* as the ability to do work. In this definition, *work* means moving an object across a measurable distance. In order to move an object, or do work, someone or something must exert a *force* against the object. A *joule* is the amount of energy required to move an object over a specific distance, such as when you lift a cellphone 1 meter or 1 yard off the ground.

Not familiar with the joule? You can also think of energy in terms of watts when you use electricity or calories when you eat food. For instance, the power company measures your energy usage in watts. A *watt* is the number of joules per second used to do the work of running all the appliances in your home. Light bulbs (and other electrical appliances) often have a wattage label, such as 60 watts or 100 watts. This measure tells you how much energy the light bulb consumes per second. Thus, a 100-watt light bulb uses more energy per second than a 60-watt bulb.

Before energy was defined in a scientifically standard way, people used the term *calorie* to mean the same thing as a joule. But these days, a calorie is most often used specifically to measure food energy. One way to understand how calories measure food energy is to think about human energy in terms of watts. The average human being emits 100 watts of energy per second (approximately 0.02 calories per second). To replace this energy, the average person needs to consume 2,000 calories each day.

Converting the Sun's Energy to a Useable Form

Environmental scientists recognize that the fundamental source of energy for most life on Earth is the sun. Stand in the sunshine on a cool fall day, and you'll feel its warmth. The sun's energy comes to Earth as waves of radiation, which you experience as heat and light. While you feel the heat, plants capture the light and convert it into chemical potential energy. Plants then store the potential energy in the form of *biomass* (biological matter that fuels nearly every animal on Earth). Only after an animal eats the plant is the potential energy from the sunlight released as kinetic energy for movement and growth.

I describe in detail how energy flows through ecosystems in Chapter 6. In this section, I focus on the processes of photosynthesis and respiration, which transform energy from the sun into energy that living organisms can use.

Spinning sugar from sunlight: Photosynthesis

Photosynthesis is the chemical process by which green plants convert sunlight into sugar. In essence, this process transforms a wave of light energy into chemical potential energy, which the plant then stores in the molecular bonds of sugar molecules.

The following steps walk you through the photosynthesis process that occurs each time the sun's light reaches the leaves of a plant:

1. Inside the *chloroplast* (a special organelle within a plant cell), a molecule of chlorophyll absorbs the light.

   The chlorophyll compound also gives the plant its green color. See Chapter 3 for details on cells and biological matter.

2. A sequence of chemical reactions transfers the sun's light energy into the chemical bonds that hold together special, energy-carrying molecules (the most common of which are called *ATP*).

   At this point, the energy originating from the sunlight is being stored in the ATP molecules as chemical potential energy.

3. The plant uses the stored chemical energy of ATP to make glucose from carbon dioxide. The plant then uses the glucose to make even larger compounds of cellulose and starch, which store energy.

   As the plant binds molecules into larger and larger chains, it captures and stores energy in the bonds to be released later. The plant uses the largest molecules to construct cell walls as the plant grows larger.

The following equation sums up the photosynthesis reaction:

$$\text{Sunlight} + 6H_2O + 6CO_2 \rightarrow C_6H_{12}O_6 + 6O_2$$

In words, this equation states that sunlight, combined with six molecules of water (H_2O) and six molecules of carbon dioxide (CO_2), produces one molecule of sugar ($C_6H_{12}O_6$) and six molecules of oxygen gas (O_2).

Through this process, green plants capture energy from the sun, use some of it to function and grow, and store some of it in their plant structure, where it's available to other organisms when they eat the plants. At the same time, the plants release oxygen into the atmosphere.

According to NASA, of all the sunlight that reaches Earth, only about 48 percent of it hits the surface and only a portion of that is captured through photosynthesis. The rest is reflected back by Earth's atmosphere or absorbed by the atmosphere.

Photosynthesis is the first stage of energy flow through an ecosystem. Check out Chapter 6 for more details on how energy flows from plants to animals and up the food chain.

Waiting to exhale: Respiration

You and all other animals on Earth rely on the energy that plants store for life. But animals aren't the only organisms that burn energy. Plants burn energy as they grow, too. In both plants and animals, the process of *respiration* — which releases stored energy for use — occurs in the mitochondria inside each cell (see Chapter 3 for the skinny on cell structure and mitochondria).

Chemically speaking, respiration is photosynthesis in reverse, as you can see in this equation:

$$C_6H_{12}O_6 + 6O_2 \rightarrow 6H_2O + 6CO_2 + energy$$

Respiration consists of a complicated series of chemical reactions. In the first stage, glucose is oxidized (see Chapter 3 for details on oxidation reactions), and the chemical potential energy of its bonds is transferred to the chemical potential bonds of an ATP molecule. The ATP molecule can then be transported throughout the cell where its stored energy is used to complete various tasks within the cell. This process releases carbon dioxide gas and water.

Respiration occurs in your cells and is fueled by the oxygen you inhale. The carbon dioxide gas you exhale is the result of a completed cycle of cellular respiration.

Only plants can photosynthesize, but both plants and animals depend on respiration to release the chemical potential energy originally captured through photosynthesis.

Figure 4-2 illustrates how closely photosynthesis and respiration are linked. As you can see, thanks to these two life-sustaining processes, plants and animals depend on each other to survive.

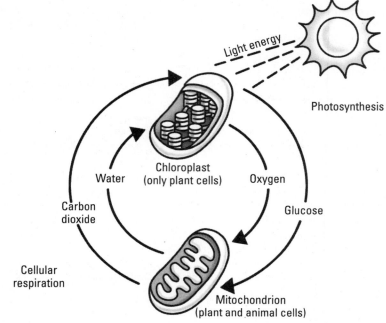

Figure 4-2:
Photosyn-
thesis and
respiration.

Illustration by Lisa Reed

Part II

Planting the Seed: Foundational Concepts in Environmental Science

The 5th Wave   By Rich Tennant

WHILE VISITING THE ISLAND OF MAURITIUS, JAN AND IRV PLOTKIN COME ACROSS THE LAST REMAINING NATURE SIGN FOR THE NOW EXTINCT DODO.

RICHTENNANT

THE DODO

Get all of it, Irv.

"Ooo! This would make a wonderful souvenir."

In this part . . .

*E*nvironmental scientists look at both the big picture and the small picture. In the big picture, communities of living things (plants and animals) around the globe correspond to specific ranges of temperature and precipitation, called *biomes.* In the small picture, each and every organism is part of an ecosystem, where it interacts with other living organisms and its nonliving physical surroundings.

In this part, you gain a better understanding of these concepts, as well as the basic principles of how organisms compete, cooperate, and coexist with one another.

Chapter 5

Studying the Environment and Your Place in It

The term *environment* refers to all the living and nonliving things that interact in a specific area. When studying a given environment, scientists often define it around a specific living thing or group of organisms, such as people. For example, consider a fishbowl on your kitchen counter. The environment of the fish includes the obvious features — the water, rocks, plants, and any microscopic organisms that live in the fishbowl — but it also includes conditions, such as temperature and light, that come from beyond the enclosed bowl. In other words, the tiny fishbowl environment includes the larger environment of your home.

The complexity of an environment depends on the size of the area being studied. The largest and most complex environment that scientists study is planet Earth, including the atmosphere that surrounds it. Like the fishbowl, the Earth's environment is influenced by factors beyond it, such as the sun. However, for the most part, the Earth is a *closed system,* which means that the matter on Earth doesn't leave the planet (see Chapter 3 for details on matter and Chapter 6 for details on closed systems).

As you explore *environmental science* (the study of the environment using a methodical approach), you need to recognize that *environment* doesn't simply mean trees, grass, and weather. Many of the environments that humans and other creatures experience today are manmade environments — such as your home and the urban landscape of a city — and these environments affect living organisms just as much as the natural environment.

In this chapter, I describe the various disciplines that contribute knowledge to environmental study, and I explain a few of the most common scientific tools that are specific to environmental science. To help you understand just how different environmental science is from environmentalism, I walk you through a brief history of the dominant ideas concerning the relationship between people and the environment. Finally, I cover the overall goal of environmental science — sustainability — and explain how everyone should be interested in reaching this goal.

Integrating Multiple Disciplines

One of the most exciting and challenging aspects of environmental science is the complexity of environmental interactions. Understanding the interactions between physical (nonliving) and biological (living) elements, as well as the many social and manmade elements of the modern world, requires knowledge too vast for one *discipline,* or subject of study. As a result, environmental science combines several different scientific disciplines, the most important of which are

- **Biology:** The study of living organisms from a microscopic level (microbiology) to the study of entire populations (population biology) and everything in between
- **Ecology:** The study of how living things interact with one another and with their physical environment
- **Toxicology:** The study of how dangerous substances (poisons) affect organisms
- **Chemistry:** The study of how the atoms and molecules of elements interact to form the living and nonliving components of the world
- **Geology:** The study of the nonliving parts of Earth, including rocks and soil, as well as the physical landscape (mountains, rivers, and so on)
- **Atmospheric science:** The study of air and water circulation above the Earth's surface, including patterns of weather and climate
- **Oceanography:** The study of Earth's oceans, including the nonliving features, such as water circulation, as well as the plants and animals that live in the oceans

Along with traditional sciences like the ones in the preceding list, social sciences, which study specific parts of human society and social interactions, also play a part in environmental science. These social sciences include

- **Economics:** The analysis of how goods are produced and distributed to meet society's needs

  ✔ **Sociology:** The study of how groups of human beings interact with one another

  ✔ **Demography:** The study of human populations, particularly their size, density, and distribution around the globe

Environmental science is often used to develop solutions to problems that are as complex and complicated as the environment itself. By combining the knowledge of so many disciplines, environmental scientists come closer to providing solutions that meet the needs of both the nonliving and living parts of the environment — including people.

Applying Common Tools of Environmental Science

Environmental scientists use a handful of tools or procedures to study, describe, and understand problems in the environment. While they use some tools that are common to the disciplines listed in the preceding section, they also incorporate a few tools of their own to help them apply the information from other disciplines in a way that's meaningful for studying the environment. The following sections take a look at three of those tools.

Reviewing case studies

A *case study* is a very useful research tool in which scientists observe a specific interaction in the complicated real-world context where it exists (rather than in the controlled, less complex context of a laboratory). Case studies help scientists build and test new hypotheses about the interactions being studied, and they usually take place over a long period of time (weeks, years, or even decades).

In the classroom, professors and textbooks often present case studies as real-world examples to illustrate a concept or problem. Good case studies link the abstract concepts of environmental interactions to what actually happens in the real world. Throughout this book, I use case studies to help you understand various environmental science concepts.

Assessing environmental indicators

Most environmental scientists are concerned with maintaining a healthy, well-functioning environment. By *well-functioning,* I mean that all the living organisms in an area are getting their basic needs (like food, water, and shelter) met.

Sometimes the needs of one organism (most often humans) negatively impact the ability of other organisms to exist. If you think that human needs clearly come first, keep in mind that healthy human societies require the food, water, and other resources that a well-functioning environment produces.

To measure the health of an environment, environmental scientists look at *environmental indicators,* or the vital signs of the environment. Scientists use environmental indicators in the same way that your doctor measures your heart rate and blood pressure during a checkup. Dramatic changes in your vital signs can indicate to the doctor that something in your body isn't functioning properly — in other words, that you're not healthy — and that you need some medical attention.

Similarly, scientists measure environmental indicators to highlight where problems exist in a given environment. Table 5-1 lists some of the most common environmental indicators, along with their units of measure.

Table 5-1	Environmental Indicators
Indicator	*Unit of Measure*
Human population size	Number of individuals
Carbon dioxide	Parts per million in the air
Food production	Kilograms per person or per hectare (1 hectare = approximately 2.47 acres)
Average surface temperature	Degrees Celsius
Sea level change	Millimeters
Annual precipitation	Millimeters
Diversity	Number of species
Extinction rate	Number of species per unit of time
Water quality	Concentration of pollutants
Habitat loss	Hectares cleared per year

Measuring human impacts: The ecofootprint

The *ecological footprint,* or *ecofootprint,* is a tool that environmental scientists use to estimate how each person impacts the Earth's supply of resources. The ecofootprint quantifies the resources, such as goods, services, food, water, and energy, used by a single person, based on the amount of land space needed to produce those resources. For example, to determine your ecofootprint, scientists trace the amount and type of food you eat back to the pastureland, farmland, and/or fishery space needed to produce it; they do the same thing for the other resources you use. Then they take the total land space and determine your ecofootprint in terms of the number of Earths it would take if everyone on the planet lived your lifestyle.

Figure 5-1 illustrates the concept of the ecofootprint and shows some of the categories of land used to calculate a person's impact.

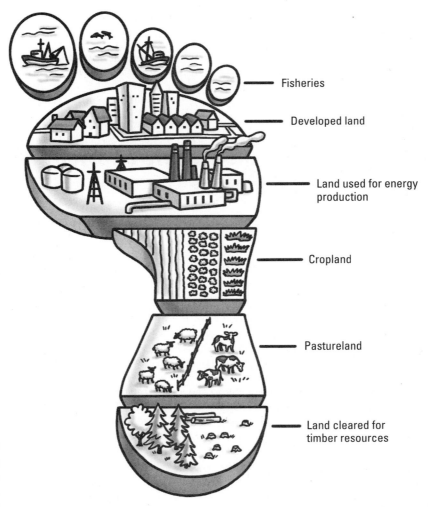

Fisheries

Developed land

Land used for energy production

Cropland

Pastureland

Land cleared for timber resources

Figure 5-1:
The eco-
footprint.

Illustration by Lisa Reed

Although the ecofootprint is only an estimate of each person's indirect use of land, environmental scientists find it to be a useful tool for quantifying and describing how growing human populations will impact the limited resources available on Earth.

To estimate your own ecofootprint, go to the following website: www.my footprint.org. To find your footprint, click on the Go! Button and answer the questions as accurately as you can. At the end, your estimated ecofootprint will be quantified as the number of Earths that would be necessary if everyone lived your lifestyle. If you've never thought about the land needed to supply your daily food, water, and other resources, you may be very surprised to see how large your footprint is!

Note: A similar but slightly different measure that you may have heard about is the *carbon footprint*. Your carbon footprint is a measure of how much carbon dioxide gas is produced to generate the goods and services (including transportation) you use. I explain why carbon footprints are important in Chapter 19.

Tracing the Development of Modern Environmental Ethics

People have been exploring the relationship between humans and the environment since ancient times; however, the dominant ideas from the last 150 years (which I describe in the sections that follow) are the ones that have shaped modern environmental thinking, including many aspects of environmental science.

Don't confuse environmental science with environmentalism. *Environmentalism* is a social and political movement with the goal of protecting the environment. In contrast, *environmental science* is the process of using a scientific approach (known as the scientific method) to study the environment (see Chapter 2 for details on the scientific method).

Environmentalists (people who are active in the environmentalism movement) and environmental scientists share a history of how people have thought about the environment over the years. In the following sections, I describe the four major stages in the development of modern environmental ethics in the U.S.

Managing for use: Utilitarian conservation

At the end of the 19th century and the beginning of the 20th century, prominent Americans (such as President Teddy Roosevelt) began to notice that what once seemed like the endless resources of the U.S. (land and timber, for example) were being wastefully destroyed.

Roosevelt and others approached the issue very pragmatically. They believed that if someone didn't properly manage the country's wilderness resources, like forests, people wouldn't get the most use out of them. Therefore, managing these resources to provide homes and jobs for as long as the resources could last was the best idea for the American people. As a result of this idea, the U.S. Government created the National Forest Service, and for the first time, forest resources were managed through rational and scientific methods rather than the free-for-all approach.

The idea that natural resources should be managed to meet the needs of the most people for the longest time is called *utilitarian conservation.* Realize that utilitarian conservation doesn't mean to save natural resources for the future but rather to use them in the present, as well as manage them to last as long as possible.

The foundational ideas of utilitarian conservation are still the basis of many resource conservation and resource management policies. Environmental scientists play an important role in helping resource managers understand how to best conserve resources so that they can be used now as well as in the future.

A life of its own: Preservation of wilderness

While Roosevelt and others sought to manage natural resources to meet the needs of people, naturalists and writers like John Muir spoke out strongly from a different perspective. Rather than conserve for use, they sought to conserve and preserve wilderness in its natural state.

Preservation of wilderness or natural resources means leaving them alone completely. The foundation of this environmental idea is that wild plants, animals, and habitats have a right to their own existence and don't exist solely to provide homes and jobs for humans.

Unlike utilitarian conservation, preservation means thinking of other organisms as being equal to humans in their right to exist and inhabit the Earth. Preservation is a fundamental goal of environmentalism today. Many environmental scientists also recognize the importance of preserving species and ecosystems, particularly to protect biodiversity, which I describe in Chapter 12.

Reducing pollution: Battling pesticides and other toxins

In the mid-20th century, following World War II, the U.S. was caught up in technological and industrial advancements that resulted in the release of massive amounts of chemicals and other toxins into the environment. At the time, people thought only about the immediate benefits the chemicals provided (such as killing mosquitoes). No one considered the long-term effects those chemicals might have on the environment (and its inhabitants).

Scientist and writer Rachel Carson put an end to this naiveté when she published the book *Silent Spring* about how pesticides were killing not only pests like mosquitoes but also beneficial insects, birds, and other wildlife. Carson's book pointed out that if the chemicals being sprayed in suburbs and farmlands were killing these animals, they might also be dangerous for humans.

The manufacturers of chemical products and many politicians resisted Carson's call to think more critically before polluting the environment. But as environmental science studies began to document the effects of toxins on animals, plants, and humans, they confirmed Carson's warnings.

Today many environmental scientists devote their careers to understanding how pollutants and toxins in the environment affect natural ecosystems and human health. Because pollution is such a big part of environmental science, I cover this topic much more extensively in Part IV.

It's a small world after all: Global environmentalism

Modern environmental thinking includes the three ideas I cover in the preceding sections (managing resources for human use, preserving wilderness, and reducing pollutants), as well as one more — thinking about the environment in terms of the entire planet, not just one country. Thus, the most current way of thinking about humans in relation to their environment is called *global environmentalism*.

The core idea of global environmentalism is that planet Earth is one large, interconnected ecosystem that the human population must share, protect, and use wisely (see Chapters 6 and 7 for details on ecosystems). In this way, global environmentalism incorporates the knowledge gained from environmental science into ideas about what is best or most sustainable for the planet.

The Overall Goal: Seeking Sustainability

Sustainability is the ability of something to persist or endure. Within the context of environmental science, *sustainability* describes a situation in which the environment provides resources while being maintained in a healthy, long-lasting state. Because people depend on the resources of the environment for life, society needs to develop sustainable ways of using those resources so that they don't disappear.

Environmental scientists seek to understand the intricate details of how humans and the environment interact so that they can help develop and implement more sustainable methods of resource use. Ideally, maintaining the environment's ability to sustain itself will allow it to continue to supply the resources that humans need to persist and endure.

Related to sustainability is *sustainable development,* which is the idea that human societies can continue to modernize in ways that minimize damage to the environment around them. Some industrialized nations learned the lesson of sustainable development the hard way when, in the process of developing to their current industrialized state, they destroyed natural resources and polluted their environments.

Reading the environment

If you're interested in the development of environmentalism and environmental science, you may want to check out some of the following books that have played an important role in changing how people think about their environment:

✔ *Walden* by Henry David Thoreau (1854)

✔ *Man and Nature* by George Perkins Marsh (1864)

✔ *My First Summer in the Sierra* by John Muir (1911)

✔ *The Fight for Conservation* by Gifford Pinchot (1910)

✔ *A Sand County Almanac* by Aldo Leopold (1949)

✔ *Silent Spring* by Rachel Carson (1962)

✔ *The Population Bomb* by Paul R. Ehrlich (1968)

✔ *Pilgrim at Tinker Creek* by Annie Dillard (1974)

✔ *Biophilia* by E. O. Wilson (1984)

✔ *The End of Nature* by Bill McKibben (1989)

✔ *Monster of God* by David Quammen (2003)

✔ *The Omnivore's Dilemma* by Michael Pollan (2006)

You may see and hear the word *sustainable* used to describe urbanized regions, such as U.S. cities. In these places, *sustainability* means implementing new strategies to maximize the efficiency of the urban landscape and to minimize the use of resources. For example, green roofs on city buildings help to naturally control rainwater runoff and interior temperatures. I describe some approaches to sustainable development and urban sustainability in more detail in Chapter 10.

To Be or Not to Be a Tree-Hugger: Looking Past Political Debates

Discussions of the environment — how to protect it, how to implement (and pay for) sustainability, whose rights should come first, and other debates — inevitably become very heated. And because changes in policy and laws are the most common way that large-scale sustainable measures are implemented, you may think you have to choose a side politically to even think about the environment.

Regardless of your political or cultural beliefs, the facts are the same:

- Billions of people have only one planet to share.
- All human beings depend on the natural resources of the Earth to survive.

Therefore, you have to move past political ideologies to objectively study the environment. After all, environmental science is for everyone who wants clean water to drink, fresh food to eat, and a safe and healthy landscape to live in, not just so-called tree-huggers or environmentalists.

Of course, your personal values will shape what you think is the best way to reach the goals of clean water, fresh food, and so on. But only after you have all the facts from environmental scientists can you make truly informed choices or vote toward the solution you think is best.

Rest assured, you don't have to ask yourself whether or not you want to be a tree-hugger in order to study the environment. Simply being human is qualification enough!

Changing perspectives through time

In the U.S., you can trace the development of environmental thinking from utilitarian conservation through modern global environmentalism along a timeline of important legislation and other social and cultural milestones.

During the early period, *utilitarian conservation* (that is, using natural resources now to meet the needs of the public) took priority over wilderness preservation. Although the first national and state parks were created during this period, they were managed to provide resources, not to be preserved as most national and state parks are today. Here's a quick look at this early period by year:

- **1849:** The U.S. Department of the Interior was created to officially manage public lands across the nation.

- **1862:** The Homestead Act encouraged Western expansion by offering free titles to unsettled land.

- **1864:** Under President Lincoln, the first U.S. state park was created in Yosemite Valley, California.

- **1872:** President Grant created the first U.S. national park at Yellowstone, Wyoming.

- **1890:** Yosemite National Park was established.

- **1905:** The National Forest Service was created to focus on resource management. It was headed by forester Gifford Pinchot.

After World War II, the U.S. found new uses for the chemicals and technology of warfare, leading scientists, advocates, and eventually the public to ask questions about how toxins and pollutants would affect the environment and human health in the future. By 1980, a number of important laws had been passed to clean up

pollution and protect the environment. Here's a look at some of the biggest milestones that took place between the 1920s and 1980:

- **1925:** Toxicologist Alice Hamilton published *Industrial Poisons in the United States,* raising concerns about the safety of industrial workers due to pollutants.

- **1949:** Aldo Leopold's *A Sand County Almanac* introduced the general public to natural science.

- **1962:** Rachel Carson's *Silent Spring* exposed the dangers of DDT and inspired citizens to call for legislation to protect the environment.

- **1963:** The first Clean Air Act was passed (later updated in 1970, 1977, and 1990).

- **1964:** The Wilderness Preservation Act created a program for protecting certain areas of wilderness from resource extraction and development.

- **1966:** The Endangered Species Preservation Act was passed.

- **1968:** The National Wild and Scenic Rivers Act protected free-flowing rivers from development.

- **1970:** The first Earth Day was celebrated on April 22.

- **1972:** The Clean Water Act was passed.

- **1975:** Congress set up the first regulations on automobile emissions, leading to the addition of catalytic converters.

- **1980:** Congress established the Superfund Act (also called the Comprehensive Environmental Response, Compensation, and Liability Act) to finance the cleanup of toxic waste at sites across the nation.

(continued)

(continued)

By the late 1980s, protecting the environment had become an international concern. Working together, the international community began to set and achieve goals for environmental conservation and protection, as well as promotion of human health. Here's a look at what's been done in the recent past:

- ✔ **1987:** The Montreal Protocol was passed, creating an international agreement to curb ozone damage by pollutants.

- ✔ **1997:** The Kyoto Protocol was passed, creating an international agreement to reduce greenhouse gas emissions that lead to global warming. By 2011, 191 countries, not including the U.S., had signed the agreement.

- ✔ **2005:** The first Millennium Ecosystem Assessment was conducted to study the consequences of ecosystem damage on human welfare.

Chapter 6

Exploring Ecosystems

· ·

· ·

*T*o study something as complex as the interaction between all living things and their environment, environmental scientists start small. They break the larger *system* (or unit) — the Earth — into smaller pieces that still have the characteristics and interactions of the larger system. These smaller pieces are called ecosystems.

An *ecosystem* includes living organisms and the environment that they inhabit and depend on for resources. Like any system, an ecosystem functions in scientifically predictable ways. In this chapter, I describe systems in general and the ecosystem specifically. In particular, I explain how matter cycles and energy flows through an ecosystem, and I describe the role that living and nonliving parts of the ecosystem play in these processes (see Chapter 3 for details on matter and Chapter 4 for details on energy).

Understanding Systems

Generally speaking, a *system* is a unit built of smaller components that work together. A bicycle or car is a system. So is a family or workplace.

Like everything around you, the components that make up a system are made up of matter (see Chapter 3 for the skinny on matter). The three types of matter that you find in systems are

▶ **Input:** Matter that enters a system

▶ **Output:** Matter that exits a system

▶ **Throughput:** Matter that flows through a system

Whether or not a system allows matter to enter, exit, or flow through it determines whether the system is open or closed. The addition of input from outside the system changes how the parts of a system interact with one another and may result in output that leaves the system.

Defining open and closed systems

A system that interacts with matter originating outside of itself is called an *open system*. In other words, an open system includes input, output, and throughput. Figure 6-1 illustrates the basic concept of an open system.

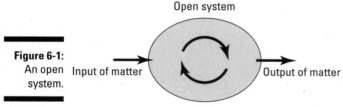

Figure 6-1: An open system.

Illustration by Wiley, Composition Services Graphics

An example of an open system is an automobile. Gasoline (input) enters the automobile when you fill up at the pump. It helps run the automobile's system of parts and then exits as exhaust fumes (output) out the tailpipe. In this example, the gasoline flows through the automobile system — entering and then exiting — and is considered throughput.

In contrast, a *closed system* doesn't allow any matter to come into or leave the system. Instead, all the matter within the closed system is recycled and used repeatedly. Figure 6-2 illustrates this concept.

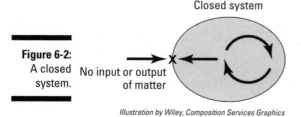

Figure 6-2: A closed system.

Illustration by Wiley, Composition Services Graphics

An example of a closed system is a snow globe. The water and "snowflakes" are sealed within the glass globe. To make it snow inside the globe, you use your hand to shake up the globe. This action transfers energy into the system but not matter; the globe stays sealed (see Chapter 4 for details on energy transfer). Similarly, you can consider the Earth a closed system, through which energy flows (from the sun) but in which matter stays and is recycled.

Energy flows through both open and closed systems, but matter enters and/or exits only open systems. (A system in which no matter or energy enters or exits is called an *isolated system* and is very rare in the natural world.)

Mapping out system dynamics

Within a system, multiple parts work together, exchanging energy and matter. Scientists who study system interactions, or *system dynamics,* have defined a few important patterns to these interactions. I describe those patterns in the following sections.

Seeking a steady state

Some systems exist in a *steady state,* or *homeostasis.* In steady-state systems, the amount of input and the amount of output are equal. In other words, any matter entering the system is equivalent to the matter exiting the system.

Some lakes exist as steady-state systems in terms of their water volume. For example, a lake that has a stream feeding water into it may also be losing water that soaks into the ground or exits by another stream. In this way, even though the stream provides a constant input of water into the lake, the lake also experiences a constant and equal output of water. As a result, the total amount of water within the lake stays the same.

Most systems continually shift inputs and outputs to maintain a steady state. Your body temperature, which remains fairly constant, is one example. When your body gets too hot, it releases heat through sweating. When your body gets too cold, it generates more heat through shivering. In this way, your body attempts to keep your temperature at a steady state by making minor adjustments to its energy inputs and outputs.

Like your body temperature, many natural systems respond to inputs by adjusting outputs. In fact, maintaining a steady state without change is difficult (and rare). So as systems try to reach equilibrium, they constantly shift inputs and outputs.

Making adjustments with positive and negative feedbacks

The adjustments that a system makes as inputs enter or outputs exit are called *feedbacks*. The two types of feedbacks are

- **Negative feedbacks:** These feedbacks slow down or suppress changes, sometimes helping the system return to a steady state.

- **Positive feedbacks:** These feedbacks lead to increased change, sending the system farther away from a steady state.

Feedbacks often set off a chain of changes, called a *feedback loop,* in the system. For example, the internal regulation of your body temperature is a negative feedback loop. A change in your body temperature triggers parts of the system (your body) to respond by increasing (shivering) or decreasing (sweating) the temperature and sending it back toward a steady state, thus suppressing change.

On the other hand, population growth can create a positive feedback loop. When more births occur, the next generation has more people to have more babies. In time, these babies grow up to have more babies, who grow up to have more babies, and so on. Thus, positive feedback loops can lead to *runaway effects* — sending a system far from its steady state.

In the context of systems, the terms *positive* and *negative* don't mean good and bad. In fact, positive feedbacks are often more dangerous than negative feedbacks because they move a system farther from stability.

Systematically Understanding the Earth's Environment

The Earth and its environment exhibit characteristics of both open and closed systems. But in terms of matter, Earth is basically a closed system in that all of Earth's matter stays put and no new matter can be added (except for the occasional meteorite!). The matter on Earth moves through different environments in cycles. In this section, I describe a few of Earth's most important cycles of matter through the environment.

Flowing through the hydrologic cycle

The *hydrologic cycle* involves water moving from the surface (most importantly the oceans) to the atmosphere, across the land, and everywhere in between. The hydrologic cycle includes various processes that change water from solid to liquid to gas form and transport it to every corner of Earth's surface (and below). In terms of water, the Earth is a closed system, so water isn't added or removed from Earth; it's simply transformed, transported, and recycled.

Since the hydrologic cycle has no beginning or end (hence the term *cycle*), you can jump in at any stage. I like to start my explanation at the oceans, where most of Earth's water is stored. Figure 6-3 illustrates the major steps of the hydrologic cycle that I outline here.

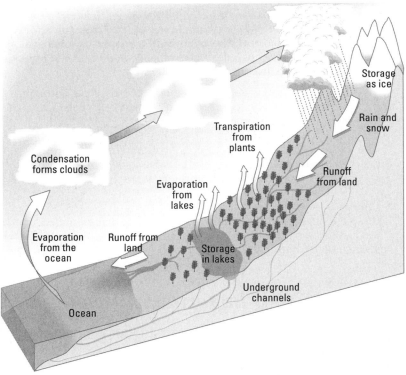

Figure 6-3:
The hydrologic cycle.

Illustration by Wiley, Composition Services Graphics

✔ **Water in the oceans moves to the atmosphere through *evaporation*, a process that changes the liquid water to *vapor*, or gas.**

After the water vapor is in the atmosphere, processes of *atmospheric circulation* transport it around the globe (see Chapter 7 for details on atmospheric circulation).

✔ **As the water vapor is carried over land, the atmosphere often releases it in the form of *precipitation* (rain or snow).**

The precipitation may stay on land in the form of snow (for a year or so) or ice (for many years), or it may move across the land as rivers and streams, and some of it will evaporate back into the atmosphere.

✔ **The water on the surface of the Earth may end up in lakes for many years, be absorbed into the soil and rocks and become groundwater, or continue to flow as runoff until it reaches the ocean again.**

Groundwater is water that flows underground toward the nearest ocean.

✔ **Plants release water into the atmosphere through a process called *transpiration*. While plants lose water to the atmosphere pretty much all the time (sort of like sweating), transpiration is higher during photosynthesis, when plants release water into the atmosphere in exchange for taking in carbon dioxide.**

This exchange of water between the atmosphere and plants is a part of the hydrologic cycle that's often overlooked. (See Chapter 4 for details on photosynthesis.)

The hydrologic cycle doesn't occur in a straight line. Throughout the path I describe in the preceding list, water is being evaporated back into the atmosphere and being added to the surface as precipitation.

Understanding the movement of water through the environment and around the planet is important when you're studying issues such as freshwater resources (see Chapter 9), water pollution (see Chapter 16), and climate change (see Chapter 19).

Recognizing important nutrient cycles

Water may be the most important and visible material that moves through Earth's environmental system, but it isn't the only one you need to know about when studying the environment. Other matter that cycles through Earth's environment includes important elements, or *nutrients,* such as carbon and nitrogen. All four of the nutrients that I describe in this section are naturally occurring elements on the Earth. Each one moves through the Earth's environment between *sources,* where the element enters the environmental system on Earth's surface, and *sinks,* where the element is stored away (usually below Earth's surface in rocks).

I've included each of the following nutrient cycles in this section because human actions have in some way altered their normal rates and processes.

Carbon cycle

In the era of global climate change, you've probably heard about the carbon cycle and the fact that burning fossil fuels, such as gasoline and coal, adds carbon to the Earth's atmosphere. But the carbon cycle involves more than just burning fossil fuels. After all, *carbon* is the element that forms the basis for all living matter on Earth (see Chapter 3 for more on matter). You find it just about everywhere on Earth that living matter currently or used to exist. Figure 6-4 illustrates the major components of the carbon cycle.

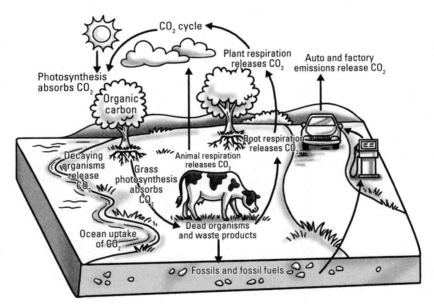

Figure 6-4: The carbon cycle.

Illustration by Lisa Reed

As you can see in Figure 6-4, a lot of the carbon in the atmosphere exists in molecules of carbon dioxide (CO_2). Green plants use the CO_2 during photosynthesis to create sugars that provide energy for living organisms. In this way, carbon cycles from the atmosphere to plants and animals. The plants and animals release some of the carbon back into the atmosphere during respiration, while some of it remains in the living organisms until they die and decompose, at which time the carbon returns to the soil. (See Chapter 4 for details on photosynthesis and respiration.)

Depending on the rate of organic decay, some of the carbon from decomposing organisms may be stored below Earth's surface for many years. (These storage places are called *carbon sinks.*) Carbon is also stored in seafloor sediments. Large amounts of carbon are found in Earth's oceans, where organisms combine it with calcium and other elements to build shells. When the animal dies, the shell remains, continuing the storage of carbon in sediments on the ocean floor.

The carbon energy sources that fuel much of modern life are the result of carbon from ancient organisms (I'm talking many millions of years ago) that never completely decomposed. Thus, the carbon was stored as coal and oil deep below the Earth's surface.

Like many natural systems on Earth, the carbon cycle maintains a relatively steady state by shifting its processes to accommodate change to input or output. For example, when the amount of carbon dioxide in the atmosphere increases, the oceans absorb more carbon dioxide, shifting the increased load of atmospheric carbon through the system as the system works toward a steady state.

Human action (that is, increased used of carbon-based fuels) has created a problem in the carbon cycle. It has sped up carbon's release into the atmosphere beyond natural rates. This increased release of carbon from below the Earth's surface without a matching increased rate of sinking carbon back into the Earth's surface overloads the atmospheric side of the system.

In response, natural processes in the carbon cycle, such as ocean absorption, work harder to take up the extra CO_2 released into the atmosphere. The result is that the oceans have become more acidic, which is bad news for organisms that have adapted to less acidic ocean waters. (I describe this problem in more detail in Chapter 19, and I cover the basics of acid-base chemistry in Chapter 3.)

Nitrogen cycle

Nitrogen is an important nutrient for building organic molecules, such as nucleic acids and amino acids (see Chapter 3). But while nitrogen is largely available in the atmosphere as gas molecules (N_2), most living things can't use nitrogen in that form. Rather, living things (specifically plants) need nitrogen that has been through a process called nitrogen fixation.

Nitrogen fixation is a chemical process through which special nitrogen-fixing bacteria transform molecules of nitrogen gas (N_2) into molecules of ammonia (NH_3). The nitrogen in ammonia dissolves into water and can be stored in soil.

In the soil, other bacteria complete a process called *nitrification,* during which ammonia interacts with oxygen (through a chemical reaction called *oxidation;* see Chapter 3) and becomes nitrite and nitrate. Plants then absorb these nitrogen compounds and use them to build organic molecules such as proteins.

Figure 6-5 illustrates some of the important components of the nitrogen cycle. Throughout the cycle, nitrogen may transfer between the living and nonliving parts of the environment with the help of nitrogen-fixing bacteria and *denitrifying* bacteria (bacteria that do the opposite of nitrogen fixation).

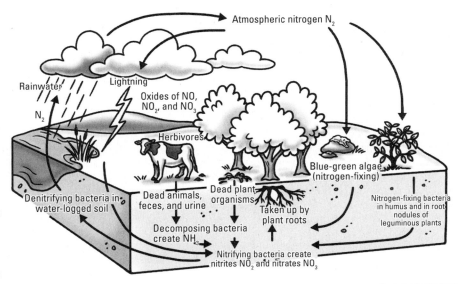

Figure 6-5: The nitrogen cycle.

Illustration by Lisa Reed

Like the carbon cycle, the nitrogen cycle experiences its share of issues thanks to human action. For example, because adding nitrogen to soil helps some crops grow, over the last few decades humans have used technology to create synthetic, or unnatural, sources of nitrogen to use as fertilizer and boost crop production. Although this is great news for food production, not all plants benefit from increased amounts of soil nitrogen. The result is that large amounts of nitrogen are being input into the soil, changing the nutrient supply in a way that encourages certain plants to grow in place of others, thus reducing biodiversity (which I explain in Chapter 12).

As they've done in the carbon cycle, humans have increased inputs of nitrogen to part of the natural nitrogen cycle without providing a way for the system as a whole to stabilize. In other words, the natural system can't keep up with the influx of nitrogen due to fertilization, so a lot of the extra nitrogen doesn't make it to the next phase of the cycle. I explain the result of this imbalance in the nitrogen cycle in Chapter 16, but here's a hint: It has to do with water pollution.

Phosphorus cycle

Living organisms use the element *phosphorus* to build nucleic acids (such as DNA) and other important organic molecules. The main source of natural phosphorus in the environment is the erosion of rocks that contain phosphorus compounds. The phosphorus eroded from rocks enters the soil and water, where it's absorbed by plants and then eaten by animals. Like other matter, phosphorus is recycled as the plants and animals decay. Over time phosphorus eventually ends up in the oceans, where it dissolves as phosphate compounds that eventually drop to the seafloor. Deep-sea sediments function as phosphorus sinks, where rocks store the phosphorus for many millions of years before it re-enters the surface part of the phosphorus cycle through uplift and erosion.

Like nitrogen, phosphorus is an important fertilizing nutrient, so humans often add it to crops to encourage growth. Humans also use phosphorus compounds in household soaps (such as laundry detergent). The phosphorus used to manufacture fertilizers and soaps comes from underground deposits of phosphorus-rich rock.

By removing the phosphorus from below ground and adding it to the Earth's surface (through mining), humans have shifted the natural phosphorus cycle, which would take many millions of years to expose the buried phosphate to the surface environment. The result is an overload of phosphorus in water systems, which leads to nutrient pollution and overgrowth of microscopic water organisms such as algae (I explain more about this issue in Chapter 16).

Figure 6-6 illustrates the phosphorus cycle, including human inputs.

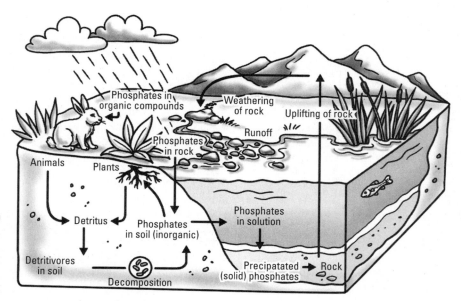

Figure 6-6:
The phosphorus cycle.

Illustration by Lisa Reed

Sulfur cycle

Another element that's important in building proteins is *sulfur*. Similar to phosphorus, sulfur is stored in rocks. Erosion and volcanic activity naturally release sulfur into the surface environment. In the natural sulfur cycle, shown in Figure 6-7, the element is distributed throughout the oceans, atmosphere, rocks, and living matter of the Earth.

Unfortunately, humans' use of fossil fuel sources, such as coal and oil, releases large amounts of sulfur that were previously stored underground into the Earth's atmosphere. The increased amounts of sulfur in the atmosphere are transferred to Earth's surface (and oceans) as acid rain. I explain the environmental problems associated with acid rain in Chapter 15.

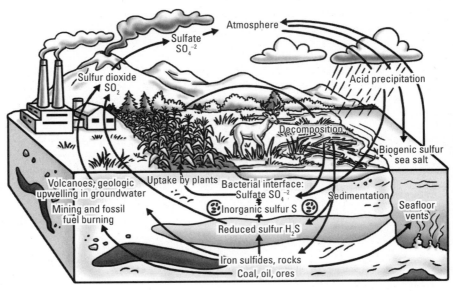

Figure 6-7: The sulfur cycle.

Illustration by Lisa Reed

Transferring Energy and Matter through an Ecosystem

To study the effects of shifting nutrient cycles on living things, environmental scientists look at ecosystems. An ecosystem can be very large, such as the entire planet Earth, very small, such as a drop of pond water, or any size in between, such as a local farm. Ecosystems are useful units to study because every ecosystem includes countless complex interactions of living and nonliving things at a level that's small enough for you to observe, record, and test hypotheses (see Chapter 2 for details on how scientists use hypotheses to answer questions).

In this section, I explain how environmental scientists define an ecosystem as a unit of study and how they study the flow of energy and cycling of matter through and within an ecosystem of any size and level of complexity.

Defining ecosystem boundaries

Considering that the entire Earth is really one big ecosystem, environmental scientists sometimes have a hard time drawing boundaries between two or more ecosystems. Occasionally, ecosystems in the natural world have clear boundaries; for example, caves are naturally bounded from too much interaction with organisms and matter outside of the cave.

More often, however, the boundaries of an ecosystem are *subjective,* meaning that scientists define ecosystem boundaries in a way that best suits the particular study. For example, if you were studying a lake and wanted to understand how the living organisms in the lake interact and use the available food resources, you might define the lake as the ecosystem, excluding anything beyond the water of the lake.

Of course, the lake ecosystem is an open system, and you have to consider the inputs of matter and energy from the land surrounding the lake, the air above the lake, and the rocks along the lake's bottom. But after you subjectively decide that the lake is the ecosystem (or the unit you will study), anything outside the ecosystem becomes either an input or an output to the system you're studying.

Getting caught in the food web

One way matter and energy move around in an ecosystem is through *consumption,* or eating. In an ecosystem, food is both matter and energy, but I focus on its role as energy here. In particular, energy enters an ecosystem through photosynthesis (see Chapter 4) and then moves from one organism to the next as one organism eats another organism. You probably know this chain of energy transfer as the *food chain.*

A food chain links one organism to another in a straight line, illustrating what eats what. Figure 6-8 shows an example of a food chain: A grasshopper eats a plant, a mouse eats the grasshopper, and an owl eats the mouse.

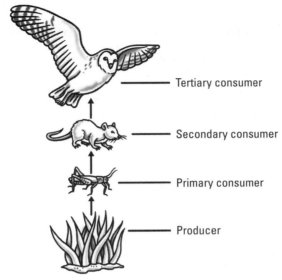

Tertiary consumer

Secondary consumer

Primary consumer

Producer

Figure 6-8:
A food
chain.

Illustration by Lisa Reed

Scientists classify the organisms in a food chain according to what position, or *trophic level*, they occupy in the chain. In a food chain, energy moves into each trophic level in sequence, as organisms consume other organisms. Scientists use the following trophic levels to classify organisms in a food chain:

✔ **Producers:** *Producers* (also called *autotrophs*) are organisms, such as green plants, that are capable of creating their own food. As producers, green plants use photosynthesis to capture energy from the sun and produce matter to store the energy. Producers are the foundation of an ecosystem, on which everything above them in the food chain depends. In Figure 6-8, the plant is the producer.

✔ **Primary consumers:** A *primary consumer* is an organism that consumes, or eats, the producers. Anytime you eat a salad, you're acting as a primary consumer. Any *herbivore*, or organism that eats plants, is a primary consumer. In Figure 6-8, the grasshopper is a primary consumer.

✔ **Secondary consumers:** A *secondary consumer* is an organism that consumes primary consumers. Secondary consumers are *carnivores*, or meat eaters, that eat herbivores. In Figure 6-8, the mouse is a secondary consumer.

✔ **Tertiary consumers:** A *tertiary consumer* is a carnivore that eats secondary consumers (*tertiary* simply means third). Tertiary consumers are often called *top predators* because they're at the top of the food chain. Other consumers seldom eat top predators. In Figure 6-8, the owl is a tertiary consumer.

Not all organisms fit neatly into a single trophic level. Some organisms eat at multiple levels. For example, grizzly bears eat both berries (making them primary consumers) and fish (making them secondary or tertiary consumers, depending on the type of fish). Similarly, insect-eating plants, such as the Venus flytrap, are both producers (they perform photosynthesis) and consumers (they eat insects).

Energy rarely moves through an ecosystem one step at a time by transferring up through individual trophic levels. More often food chains like the one illustrated in Figure 6-8 are just one of many complexly interwoven chains that create the ecosystem's food web. A *food web,* as shown in Figure 6-9, contains multiple food chains, or paths, that energy can take as it passes from one organism to another.

Figure 6-9:
A food web.

Illustration by Lisa Reed

Within a food web, scientists identify roles that go beyond the basic trophic levels that I list earlier in this section. Other roles an organism may play in a food web include

- ✓ **Scavenger:** A *scavenger* is an organism that feeds on the remains of already-dead organisms. Scavengers are carnivores because they eat herbivores, but instead of hunting and killing their own food, they show up after the herbivores have died (or been hunted and killed by something else) and help themselves to what remains.

- ✓ **Detritivore:** *Detritivores* are organisms that eat *detritus*, or dead tissue and organic waste. They break down the dead organic material into smaller pieces as a first step to recycling it. One common example of a detritivore is the dung beetle, which survives by consuming and processing the fecal waste of other organisms.

- ✓ **Decomposer:** *Decomposers* do the important work of consuming and transforming organic material back into its very basic molecules, thus completing the matter-recycling process started by detritivores. After decomposers, like bacteria and fungi (such as mushrooms), break down the organic matter, the matter is ready to be reused by other organisms in the ecosystem.

Measuring productivity

The *productivity* of an ecosystem is the amount of energy available to support the organisms in the ecosystem. To *quantify,* or measure, the energy in an ecosystem, scientists start by looking at the green plants because the total amount of energy available in an ecosystem depends on the amount of sunlight that producers capture through photosynthesis.

During photosynthesis, producers transform sunlight into *biomass,* or living matter. The total energy captured by producers is called the *gross primary productivity* of the ecosystem. But not all that energy goes out into the ecosystem. Producers have to use some of the captured energy to fuel their own living process of cellular respiration (see Chapter 4 for details). Therefore, the total amount of energy available to be transferred and used throughout the ecosystem is really the gross primary productivity minus the energy used for producers' respiration. This value is called the *net primary productivity.* Another way to look at this measure is by using the following simple equation:

Gross primary productivity – Energy for respiration = Net primary productivity

Building ecological pyramids: Illustrating energy and biomass flow

To visualize the flow of energy and organic matter through the trophic levels of an ecosystem, scientists construct pyramids.

An *energy pyramid,* such as the one illustrated in Figure 6-10, shows how the energy in an ecosystem is distributed among the trophic levels (in joules, J). The foundation of the pyramid represents the net primary productivity of the producers in the ecosystem. Scientists often use joules to measure this value (see Chapter 4 for details). At each trophic level, only about 10 percent of the energy from the previous trophic level is available. Each time one organism transfers energy to the next organism, most of the transferred energy (about 90 percent) fuels the daily living processes of the organism. As a result, less energy is leftover at the next level.

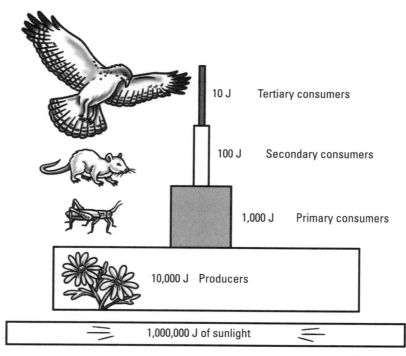

10 J   Tertiary consumers

100 J   Secondary consumers

1,000 J   Primary consumers

10,000 J   Producers

1,000,000 J of sunlight

Figure 6-10:
An energy
pyramid.

Illustration by Lisa Reed

Similar to an energy pyramid, a *biomass pyramid* illustrates the distribution of biomass in an ecosystem. Like energy, the total amount of organic matter decreases as it moves up through the trophic levels. Biomass is a way to store energy, and at each level in the biomass pyramid some of it will be used as fuel while the rest is unconsumed (and therefore left to the decomposers, which are always present).

Figure 6-11 shows an example of a biomass pyramid. One way to understand what this pyramid illustrates is to read it like this: To support one tertiary consumer (the eagle, a top predator), the ecosystem needs to contain 11 primary consumers (the fish). To support those 11 fish, the ecosystem needs to contain a huge number of producers (the photosynthetic plankton), which are actively capturing sunlight.

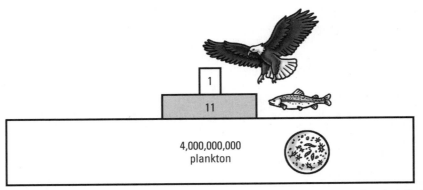

Figure 6-11:
A biomass
pyramid.

1

11

4,000,000,000
plankton

Illustration by Lisa Reed

REMEMBER

Because producers capture, convert, and store energy as biomass, energy and biomass pyramids are closely related. You may see both ideas combined and presented as a *trophic pyramid*, which represents both energy and biomass transfer in the ecosystem.

Viewing the Earth as one, big living organism

In the 1970s, a few scientists began to think about the Earth as a single ecosystem or organism of its own. The most famous of these scientists was James Lovelock, who proposed the *Gaia hypothesis,* asserting that the Earth functions as a living organism. Named after the Greek goddess Gaia, this hypothesis suggests that the Earth itself has evolved and changed over time much like a living organism.

For many years (and even today), some people think Lovelock's ideas are more related to New Age mysticism than they are to science. But the Gaia hypothesis isn't really as far out as it may seem at first. After all, scientists have been studying how the Earth's systems, including the atmosphere, oceans, and ecosystems, interact for many years. Not to mention, whole scientific disciplines (Earth system science, biogeochemistry, and global ecology, just to name a few) have been born to further study these complex interactions.

In recent decades, the Gaia hypothesis has gained support as scientists have gathered evidence that proves how closely linked the Earth's multiple systems really are and how each system is influenced by the other systems it interacts with. In fact, many scientists now consider Lovelock to be the founder of the *Gaia theory* (see Chapter 2 for insight on the difference between hypothesis and theory). In simplest terms, the Gaia theory explains how the various complex systems of the Earth (oceans, atmosphere, living things, and nonliving matter) interact with one another to keep conditions habitable for existing life. In terms of climate change, this means that if the climate gets too warm or cold, other systems on Earth will respond with feedbacks to shift the system back toward equilibrium. Over time, the Earth itself will evolve along with the living things that inhabit it.

Viewing the Earth from this perspective offers new insight into understanding how humans can affect the Earth's environment and is particularly useful in answering questions about modern climate change — how the atmosphere, oceans, and ecosystems will respond to human influences and the impact of growing human populations. (You can find more on modern climate change in Chapter 19.)

Chapter 7

Biogeography: How Earth's Climate Shapes Global Ecosystem Patterns

From the arctic tundra where grizzly bears and reindeer inhabit a treeless landscape to the dense, forested jungles where monkeys swing through the treetops, life on Earth comes in a wide variety of sizes, shapes, and combinations. But even with so much variety, life on Earth fits into several recognizable patterns around the globe. *Climate conditions* — the ranges of temperature and moisture throughout the year — determine these patterns, and like the plants and animals on Earth, these conditions vary as you move around the Earth's surface.

In this chapter, I explain a few of the factors that create the different climate conditions on Earth, and I describe how climate shapes the communities of living things — called *ecosystems* — in different parts of the world. I also outline the major ecosystem categories, or *biomes*, that scientists recognize on Earth.

Positioning Ecosystems: Latitude and Altitude

Organisms in an ecosystem are closely tied to the temperature and moisture conditions of their environment, which is why you don't find polar bears in tropical forests or amphibians in the Arctic. If you look closely at the plants and animals living on Earth's surface, you may notice that similar ecosystems occur in regions of similar climate around the world. You may also notice that the pattern of ecosystems corresponds to changes in altitude and latitude:

- **Latitude:** Lines of *latitude* indicate the distance of a location north or south of the equator. The equator circles the Earth at a point that's half-way between the North and South Poles. You've probably seen latitude lines on a map or globe; they're the lines going east and west around the globe. The other lines that go from the North Pole to the South Pole are lines of *longitude*. Scientists measure both latitude and longitude in degrees. In the case of latitude, the equator is zero latitude, and the other lines are measured as degrees north or south from the equator. Scientists use longitude to describe the east-west position of something on Earth, but lines of longitude aren't directly related to patterns of climate and ecosystems the way lines of latitude are.

  To remember the difference between latitude and longitude, remind yourself that lines of *lat*itude are *flat* and lines of *long*itude are *long*.

  When scientists talk about latitudinal positions, they may say something is in the *lower latitudes,* which means it's closer to the equator, or something is in the *upper* or *high latitudes,* which means it's near the poles of the Earth. If something is in the *mid-latitudes,* it's somewhere in between the equator and the poles.

- **Altitude:** *Altitude* is another word for *elevation,* or the height of something from a point of reference (usually sea level). Altitude increases as you move up a mountain or as your airplane takes off, moving higher into the sky. (Eventually, the pilot tells you that you have reached cruising altitude — the distance above the Earth's surface at which you'll stay during your flight.)

As latitude or altitude increases, climate conditions get cooler and ecosystems are populated by plants and animals that are best adapted to those conditions. However, latitude and altitude alone can't explain the differences in temperature and moisture around the Earth. Climate factors, such as the way in which the sun's energy strikes and is absorbed by the Earth, as well as the ocean and atmospheric circulation patterns, also play an important role. Keep reading to find out more about these factors.

Recognizing Major Climate Drivers

The fact that climate conditions are generally cooler with increasing latitude is a direct result of where the Earth is in relation to the sun and how the atmosphere and oceans move water around. In this section, I describe the most important influences on the Earth's overall climate, including differential absorption of the sun's energy, air circulation in the atmosphere (and how increasing altitude results in cooler temperatures), and ocean circulation patterns.

Bringing on the heat: Differential absorption of the sun's energy

Different locations on Earth are heated differently based on their positioning toward the sun and the way in which they absorb the sun's energy. How the sun's energy heats the Earth depends on three main factors:

✔ **Sunlight hits the Earth at different angles in different places.** Because the Earth is spherical like a ball, sunlight hits the Earth's surface in different ways. Sunlight is most direct near the equator, where it concentrates its energy on a small amount of surface area. At the mid-latitudes, sunlight has to spread out over a larger surface area, so it's slightly weaker as it moves through more of the atmosphere to reach the Earth's surface. The weakest levels of sunlight hit the Earth at the poles, where it travels through much more atmosphere and spreads out across a much larger land surface. Figure 7-1 illustrates the different angles at which sunlight hits the Earth's surface.

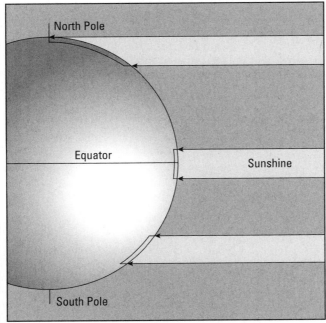

Figure 7-1:
Angles of sunlight at different latitudes.

Illustration by Wiley, Composition Services Graphics

⮫ **The Earth is tilted on an axis.** The Earth spins on its axis (which you and I experience as the transition between day and night) as it moves around the sun. However, the axis of the Earth isn't situated straight up and down (relative to the plane of rotation). Rather, it sits at an angle, causing parts of the Earth's surface to always be closer to the sun than others. Which parts are closest to the sun vary throughout the year. The parts of the Earth closest to the sun experience longer days and a more direct angle of sunlight, making it warmer.

For example, if you live in the Northern Hemisphere, you experience cold temperatures (winter) in December because the northern half of the Earth is tilted slightly away from the sun during that time of year. At the same time, people in the Southern Hemisphere experience warm temperatures (summer) because the southern half of the Earth is tilted toward the sun. By June, the Earth has orbited around to the other side of the sun, meaning that the Northern Hemisphere is now tilted slightly toward the sun. As a result, the Northern Hemisphere experiences summer, with more hours of daylight and more direct sunlight, while the Southern Hemisphere experiences winter. Figure 7-2 illustrates this concept.

⮫ **Land, water, and ice absorb heat differently.** When sunlight hits a location on Earth not covered with water or ice, most of the heat energy is absorbed and concentrated at the surface, where it becomes warm. Sunlight that hits water is partially reflected by the water's surface and partially absorbed and distributed throughout the body of water. Ice, which covers the land and water surface at the poles and on some mountaintops, absorbs very little heat, reflecting almost all the sun's energy back into the atmosphere. If ice absorbed too much heat, it would melt!

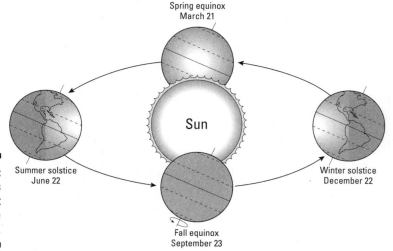

Figure 7-2: The Earth's tilt and orbit create seasons.

Illustration by Wiley, Composition Services Graphics

The wide blue yonder: Atmospheric circulation

The Earth's absorption of the sun's energy, which I discuss in the preceding section, is only part of the complex climate equation. The layer of gas surrounding Earth, known as the *atmosphere,* also plays an important role in moving heat and moisture around the Earth, and it does so through patterns of *atmospheric circulation.* Atmospheric circulation patterns include convection cells, prevailing winds, and orographic uplift.

Making it rain: Hadley cells

As heat from the sun hits Earth, it warms up the gases in the atmosphere, causing them to circulate or convect. *Convection* transfers heat from one place to another through motion (see Chapter 4 for more details).

Near the equator, or at low latitudes, the sun's energy warms the air, causing it to move upward and toward the poles. The upward motion at the equator draws in colder air along the surface of the Earth from higher latitudes (near the poles), pulling it toward the equator. This cycling of air creates a series of convection cells that scientists call *Hadley cells* (see Figure 7-3).

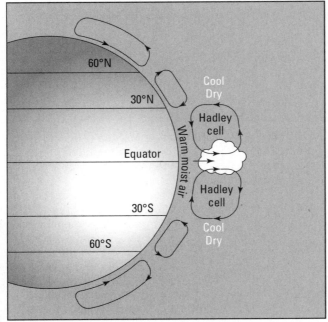

Figure 7-3:
Air circulation in Hadley cells.

Illustration by Wiley, Composition Services Graphics

On the ground, Hadley cells result in heavier rainfall near the equator. Here's why: The heat near the equator increases the evaporation of water (turning it into gas form, or water vapor). As the heated air, full of water vapor, moves upward, it cools, and the water vapor condenses into clouds and eventually rain.

Similar convection cells occur in the atmosphere near the poles, where cold air sinks to the ground, forcing air at Earth's surface to move up and away from the poles, where it warms. A third set of atmospheric convention cells occurs at the mid-latitudes.

Moving air around: Prevailing winds and the Coriolis effect

All around the Earth scientists observe fairly steady patterns of *wind* (or directions of airflow). These *prevailing winds,* as they're often called, are the result of the air being moved by convection currents (the Hadley cells at the equator and other convection cells near the poles and mid-latitudes) combined with the Earth's rotation.

As air moves away from the mid-latitudes toward the poles and toward the equator, it doesn't move in a straight line. Rather, it moves in a slightly curved direction as a result of the Earth's rotation. This phenomenon is called the *Coriolis effect.* To help you visualize how the Coriolis effect works, imagine that you're standing at the North Pole and tossing a baseball toward the equator. While the ball travels through the air, the Earth below it is rotating, so when the ball reaches the equator, it lands in a location somewhere to the west of where you were aiming. Figure 7-4 illustrates this example.

Near Earth's surface, the Coriolis effect creates wind (and water) patterns that move to the east toward the equator and to the west toward the poles. These prevailing wind patterns are responsible for moving clouds around the globe and, thus, for creating patterns of weather in different regions.

Getting in the way: Orographic effects

In addition to the gases in the atmosphere and the prevailing winds, the Earth's land surface also plays a role in atmospheric circulation. In particular, mountains change the flow of air circulation through what scientists call *orographic effects.* As altitude increases, air becomes cooler due to decreasing atmospheric pressure. When a prevailing wind moves toward a mountain range, the air is forced to move up and over the mountain. At these higher altitudes, it becomes cooler and results in rainfall as the moisture in the air condenses. As the wind moves up and over the mountain, the moisture is squeezed out of it, in a sense, so that the air is dry on the other side of the mountain. The dry side of the mountain is called the *rain shadow* and is often much drier than the windward side. Figure 7-5 illustrates what orographic effects look like.

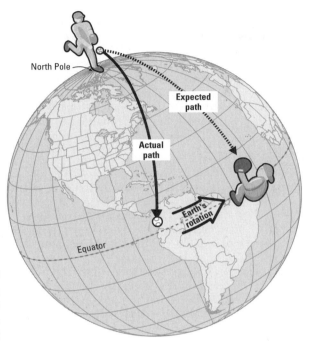

Figure 7-4:
The Coriolis effect.

Illustration by Wiley, Composition Services Graphics

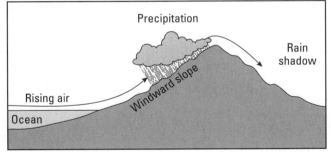

Figure 7-5:
Example of orographic effects.

Illustration by Wiley, Composition Services Graphics

Orographic effects are important for local climate and weather patterns and sometimes play a major role in the distribution of biomes, which I describe later in this chapter.

Making waves around the globe: Patterns of ocean circulation

Along with patterns of air movement in the atmosphere, the movement of water through the oceans helps determine weather and climate conditions for different regions of the world. The three main patterns of ocean circulation are gyres, upwelling, and thermohaline circulation.

Stuck in the middle: Gyres

As the prevailing winds in Earth's atmosphere blow across the surface of the oceans, the winds push water in the direction that they're blowing. As a result, the surface water of the oceans moves in concert with the air above it. This dual movement creates large circular patterns or *gyres* in each of the planet's oceans. The ocean gyres move clockwise in the Northern Hemisphere and counterclockwise in the Southern Hemisphere.

Ocean gyre circulation moves cold surface water from the poles to the equator, where the water is warmed before the gyres send it back toward the poles. The water's temperature influences the temperature of the air: Cold currents bring cooler air to the coastline as they move toward the equator, and they bring warmer air to the continents they pass on their way back toward the poles (because the currents were heated near the equator).

In Chapter 18, I discuss garbage in the oceans and explain the location of the major ocean gyres, where large amounts of trash are concentrated due to these circular water currents.

Stirring up nutrients: Upwelling

Sometimes the movement of surface currents along a coastline leads to a circulation process called *upwelling*. As a result of the Coriolis effect that I describe earlier in this chapter, upwelling commonly occurs on the west coast of continents, where the surface waters moving toward the equator are replaced by deeper cold water that moves up to the surface. The deep water brings with it nutrients from the bottom of the ocean. These nutrients support the growth of primary producers (see Chapter 6), which support the entire food web in the ocean.

Regions of the world where deep ocean upwelling occurs are often very productive with high numbers of many different types of organisms living in them.

Warm and salty: Thermohaline circulation

The largest circulation of water on the planet is a direct result of changes in temperature and salinity. *Salinity* is the measure of dissolved salt in water. The pattern of ocean currents related to salinity and temperature is called the *thermohaline circulation* (*thermo* = heat; *haline* = salt). Figure 7-6 gives you a general idea of what this pattern looks like.

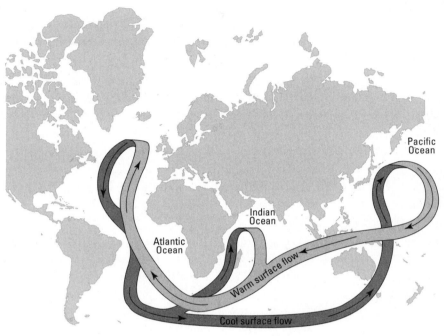

Figure 7-6:
Thermo-
haline
circulation
of ocean
water.

Illustration by Wiley, Composition Services Graphics

Sometimes called the *thermohaline conveyor belt,* this circulation pattern moves cold water around the globe in deep water currents and warmer water in surface currents. A single molecule of water being transported by thermo-haline circulation may take a thousand years to move completely throughout the Earth's oceans. The conveyor is driven by changes in the density of water as a result of changes in both temperature and salinity. Here's how this circulation pattern works:

1. Warm water in a shallow current near the surface moves toward the North Pole near Iceland. As this water reaches the colder polar region, some of it freezes or evaporates, leaving behind the salt that was dissolved in it. The resulting water is colder and has more salt per volume than it did before (and thus is more dense; I describe density in Chapter 3).

2. The cold, dense, salty water sinks deeper into the ocean and moves to the south, as far as Antarctica. After it makes its way near Antarctica, the cold, deep current splits, one branch moving up toward India into the Indian Ocean and the other continuing along Antarctica into the Pacific Ocean.

3. Each branch of the cold, deep current is eventually warmed in the Indian Ocean or the northern part of the Pacific Ocean. Although the water still contains the same amount of salt, it's a little less dense because it's warmer than the cold water surrounding it; as a result, it moves upward, becoming a surface current.

4. The warm, shallow, less dense surface current moves to the west, across the Pacific Ocean, and into the Indian Ocean, where it rejoins the Indian Ocean branch. Both branches then continue into the Atlantic Ocean and head back toward the North Pole.

Scientists who study global climate change are interested in how increased ice melting in the Arctic and Greenland will affect the thermohaline circulation. The addition of large amounts of fresh water will reduce the salinity and density and may change the pattern of global ocean circulation.

Separating the Globe into Biomes

The combined influence of the sun's energy, atmospheric circulation, and ocean circulation patterns results in environmental variation, or differences in ecosystem characteristics on Earth. Scientists organize the various ecosystems on Earth into a few broadly defined categories, called *biomes,* based on shared climate and geographic conditions. Examining the environment through biomes is useful because they connect climate conditions to the characteristics of living communities.

A biome is a region of the Earth with plant and animal communities that have adapted to a specific range of temperature and moisture conditions. Scientists define biomes by climate conditions rather than geographic locations, but because geography influences climate, biomes also illustrate geographic patterns.

You can describe biomes with a *temperature and precipitation graph* or *climate graph,* such as the one shown in Figure 7-7.

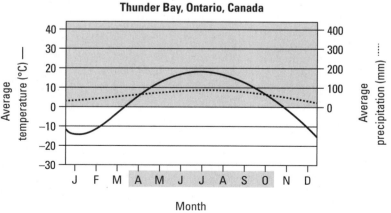

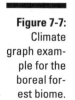

Figure 7-7: Climate graph example for the boreal forest biome.

Thunder Bay, Ontario, Canada

Month

Illustration by Wiley, Composition Services Graphics

In Figure 7-7, the left axis represents average annual temperature in degrees Celsius, the right axis represents average annual precipitation in millimeters, and the letters along the bottom of the graph indicate the months of the year. The shaded months indicate the *growing season* for the biome — in other words, the months of the year when the temperature is above freezing and plants can grow. Each biome displays a different pattern of annual temperature and precipitation across the year, which scientists can link to the patterns of plant growth and other ecosystem characteristics (such as soil nutrients). The following sections introduce you to the different biomes on Earth.

Tundra to tropics: Terrestrial biomes

Every living thing thrives under specific conditions of temperature and moisture. Humans have created wool sweaters and air conditioning so they can live anywhere, but other creatures aren't so lucky. Hence, the range of precipitation and temperature throughout the year determines the ecosystems on land, or the *terrestrial biomes*. Scientists often describe terrestrial biomes as being either temperate or tropical:

- ✔ *Temperate* biomes are located in mid-latitudes and are cooler and drier than tropical biomes (see the earlier section "Positioning Ecosystems: Latitude and Altitude" for details on mid-latitudes).

- ✔ *Tropical* biomes are located closer to the equator and are, thus, warmer and wetter in general than temperate biomes.

Figure 7-8 illustrates how changing temperature and precipitation levels combine to create some of the different biomes. Keep in mind that temperature drops with both increasing altitude (up a mountain) and increasing latitude (toward the poles).

Within the broad categories of temperate and tropical, scientists identify more specific biome categories that include vegetation characteristics in their descriptions. Here are descriptions of the most commonly recognized terrestrial biomes:

- ✔ **Tundra:** The *tundra biome* is cold, dry, and treeless. Only low-growing plants thrive in the tundra because the growing season (the warmest time of the year) is very short and the annual rain (or snow) fall is minimal. Tundra soil is usually frozen (called *permafrost*), and only the upper surface defrosts during the summer. Tundra biomes include arctic tundra in the northern parts of Russia, Canada, and Alaska, Antarctic tundra along the edges of Antarctica, and alpine tundra, which exists on any mountain top at an altitude high enough to experience cold, dry tundra conditions.

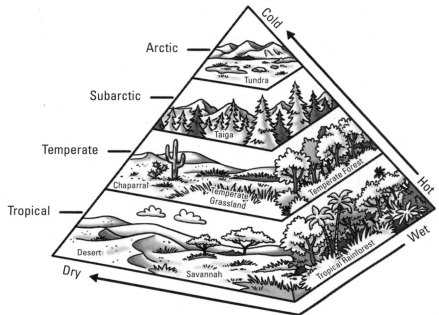

Illustration by Lisa Reed

Figure 7-8:
Some
terrestrial
biomes.

✔ **Boreal forest:** The *boreal forest biome* (also called *taiga*) consists of evergreen cone-bearing trees, such as the spruce, pine, and fir species. The climate conditions in the boreal forest are cold and relatively dry, although it gets enough precipitation each year for trees to grow. The soils in the boreal forest defrost each year, but they aren't very nutrient rich because of the slow decomposition processes during the short summer months. Most boreal forest biomes are located across the northern portions of North America, Europe, and Russia and on mountains just below the tundra biome in elevation.

✔ **Temperate grassland:** The *temperate grassland biome* is found in central North America and throughout central Europe and Asia. This biome has dry summers and very cold winters that limit plant growth. Grasses thrive, but trees and woody shrubs don't. Other names for this biome include *prairie, pampas,* and *steppe.* Humans often use temperate grassland biomes for grazing animals. The driest areas of these grasslands are too cold and dry even to support the growth of grasses and are considered *cold deserts* with little to no plant life at all.

✔ **Woodland or chaparral:** The *woodland* or *chaparral biome* is home to slow-growing plants that have adapted to occasional fire disturbance and hot, dry summers with very little precipitation (see Chapter 8 for details on fire-adapted communities). Although winters in the woodland biome aren't too cold or dry, the summer conditions of heat and drought determine which plants can survive year round. You find woodland biomes on the southern California coast, in southern South America, and around the Mediterranean. Humans often use them for animal grazing or grapevine agriculture. In other words, they're the wine-making regions of the world!

✔ **Temperate deciduous forest:** The *temperate deciduous forest biome* gets more than 39 inches of precipitation every year. All that moisture supports the growth of a variety of tree, shrub, and plant species. The trees that thrive in these biomes are *deciduous,* which means they seasonally lose their leaves (think maple, oak, and beech). Throughout the year, this biome experiences a wide range of temperatures with warm summers and cold winters. Temperate deciduous forests create highly nutritious, productive soil that humans often convert to farmland.

✔ **Temperate rainforest:** The *temperate rainforest biome* experiences mild annual temperatures (not too hot and not too cold) and abundant rainfall throughout the year. This biome is most common along coastlines, such as along the North American west coast (from southern Alaska to northern California), the southern tip of South America's west coast, and throughout New Zealand. The seaside location plays an important role in moderating temperatures throughout the year and in bringing precipitation to this biome.

✔ **Tropical rainforest:** The *tropical rainforest biome* is warm and wet throughout the year, and it's very biologically diverse thanks to those climate conditions, which support extensive plant growth (see Chapter 12 for an explanation of biodiversity). You find tropical rainforests along the equator in South America, Africa, and southeast Asia. A distinctive feature of the tropical rainforest is that the vegetation creates layers within the forest. For example, a *canopy* at the top of the forest shades everything below, and an *understory* of shade-loving trees and plants that live attached to tree trunks (called *epiphytes*) create miniature ecosystems above the forest floor.

✔ **Savannah:** The *savannah biome* is warm year round with distinct wet and dry seasons. Depending on how dry the dry seasons are, some areas of this biome (those that get more moisture) support trees and shrubs, while other areas are more open with grasslands sparsely populated by trees. Savannah biomes are located throughout much of Africa, India, and south Asia, as well as Central America and eastern South America. The soil in this biome is rich in nutrients, so humans often use it for agriculture.

✔ **Subtropical desert:** The *subtropical desert biome* is the sandy, treeless desert you often see in movies. Extreme heat, extreme drought, and very little vegetation define this biome. Plants that do live in subtropical deserts are specifically adapted to preserve moisture; examples of such plants include cacti and other *succulents*, which have tough exteriors and interiors that store water throughout the year. When it rains in the subtropical desert, the minimal plant life explodes into bloom, taking advantage of the availability of water to grow flowers, spread pollen, and reproduce quickly. Some examples of subtropical deserts are the Mojave Desert in southeastern California, the Sahara Desert in Africa, and the Arabian Desert in Saudi Arabia.

Table 7-1 summarizes the characteristics of Earth's terrestrial biomes.

Table 7-1	Terrestrial Biome Characteristics				
Biome	*Plant Life*	*Annual Precipitation*	*Annual Temperature*	*Soil Quality*	*Uses*
Tundra	Low-growing plants, no trees	Low	Cold	Frozen	
Boreal forest	Evergreen trees	Low	Cold	Poor	Logging
Temperate grassland	Grasses	Low	Hot summer, cold winter	Highly productive	Grazing, farming
Woodland/chaparral	Woody shrubs, oak	Low	Hot summer, cool winter	Low in nutrients	Grazing, grape agriculture
Temperate deciduous forest	Deciduous trees	High	Warm summer, cold winter	Rich in nutrients	Agriculture
Temperate rainforest	Large conifers, ferns	High	Mild year round	Low in nutrients	Logging
Tropical rainforest	Large tropical trees	High	Warm year round	Low in nutrients	Logging, agriculture
Savannah	Grasses, shrubs, trees	Low	Warm	Rich in nutrients	Agriculture, grazing
Subtropical desert	Cacti, other succulents	Very low	Hot	Very poor	

Depending on depth: Aquatic biomes

In addition to terrestrial biomes, much of the Earth is covered by *aquatic biomes,* or water-based ecosystems. Similar to terrestrial biomes, aquatic biomes are defined in part by temperature. Unlike terrestrial biomes, however, aquatic biomes are strongly influenced by depth, or the distance they are from sunlight.

Freshwater biomes are aquatic ecosystems found on the continents, where the water isn't salty. They include

- ✔ **Streams and rivers:** *Stream and river biomes* have flowing fresh water. The plant and animal communities in these biomes change depending on how fast the water is flowing and how much dissolved oxygen is present in the water (see Chapter 16 for details on the importance of dissolved oxygen in water).

- ✔ **Lakes and ponds:** *Lake and pond biomes* are small to large bodies of unmoving (or still) fresh water. Within a lake or pond, the ecosystem changes as the distance from the shore and the surface changes. The *littoral zone* around the edges of the lake or pond is often rich in plant life with high amounts of sunlight and shallow water. The littoral zone is home to amphibians, reptiles, and snails, as well as fish. Surface waters in the middle of a lake or pond are part of the *limnetic zone.* The limnetic zone continues in depth as long as there's enough sunlight for photosynthesis. Lakes or ponds that are deep enough may also have a *profundal* or *benthic zone,* which doesn't receive much sunlight. The profundal zone is home to decomposers and detritivores (see Chapter 6).

- ✔ **Wetlands:** *Wetland biomes* occur anywhere that the land surface is saturated or submerged underwater for at least part of the year. Scientists define wetlands by the type of plant life they support. Wetlands with trees are called *swamps,* while wetlands with communities of reeds and cattails rather than trees are called *marshes.* Wetlands are highly productive. They also provide important ecosystem services such as filtering rainwater and buffering the effect of flooding on nearby terrestrial ecosystems. (See Chapter 16 where I describe how wetlands are used to clean wastewater.)

Aquatic biomes called *estuaries* straddle the regions between fresh water and salt water. Estuarine biomes include those biomes along coastlines and in shallow seas:

- ✔ **Salt marshes:** *Salt marshes* are located where rivers enter a body of salt water. These grassy aquatic biomes are highly productive and provide an important habitat for fish. They're most common in temperate regions.

- ✔ **Mangrove swamps:** *Mangrove swamps* are found along coastlines in more tropical regions of the world. The trees that grow in mangrove swamps are uniquely adapted to salt water. The roots of these trees stabilize the coastline by holding sediment in place, and in doing so, they create an important habitat for fish and birds.

Swimming out to sea: Ocean ecosystems

While temperature and moisture conditions define the biomes on land, distance from the shore and depth from the surface define biomes in the oceans. The following terms describe specific ecosystems found along coastlines and in shallow waters. Note that ocean scientists, called *oceanographers*, use the term *zone* rather than biome when describing ecosystems of the ocean.

- **Intertidal zone:** Along coastlines, the *intertidal zone* — the area covered and uncovered by the tides each day — is home to a unique ecosystem. Organisms that live in the intertidal zone must adapt (twice a day!) to extreme changes in water as the tide moves in and out. The intertidal zone also experiences the activity of waves crashing on the shore. To adapt to these conditions, creatures like barnacles, mussels, crabs, and sea stars have found ways to conserve water, attach themselves to rocks, and burrow into the sand.

- **Coral reefs:** Farther out into the ocean, especially in warm, shallow, tropical zones near the equator, are *coral reefs*. Coral reefs are the most diverse aquatic ecosystem, supporting a huge variety of plants and animals. The reef itself is built from the shells of tiny coral animals, most of whom live in apartment-like structures, one animal per hole of the coral structure. The health of a coral reef ecosystem depends on temperature, access to sunlight, and pH of the ocean waters. These biomes are in danger from climate warming, which I describe in Chapter 19.

Far from shore, beyond the edge of the continental shelf in the open sea, the water becomes much deeper. Ecosystems in the open sea change depending on the amount of sunlight that penetrates the water. Here are the three most commonly described zones in the open sea, based on depth and access to sunlight:

- **Pelagic zone:** The *pelagic zone* is the upper layer of open seawater. The place where the most sunlight penetrates (to a depth of about 200 meters or 650 feet) is called the *euphotic zone;* it's home to the ocean's producers (particularly plankton). Animals in this zone include mammals (whales and dolphins), as well as fish. Deeper waters in the pelagic zone (below 200 meters) are part of the *bathyal zone,* which receives less sunlight than the euphotic zone above it and is sometimes called the *disphotic zone.*

- **Benthic zone:** The bathyal zone of the pelagic zone extends to the *benthic zone,* found at the seafloor, which includes the sand, silt, and other sediments as well as the creatures that live there. The benthic ecosystem gets its nutrients from the pelagic zone above it, particularly dead organisms that fall to the seafloor. Organisms in the benthic zone include seaweed, fish, fungi, and bacteria, as well as worms, sea stars, and other invertebrates.

🗸 **Abyssal zone:** No sunlight at all reaches the darkest depths of the ocean, which is called the *abyssal zone* or *aphotic zone*. The organisms that live in the abyssal zone are adapted to the absence of sunlight, and some even create energy through processes of *chemosynthesis* — transforming chemicals into energy — rather than photosynthesis.

Figure 7-9 illustrates the different ocean zones in the open sea.

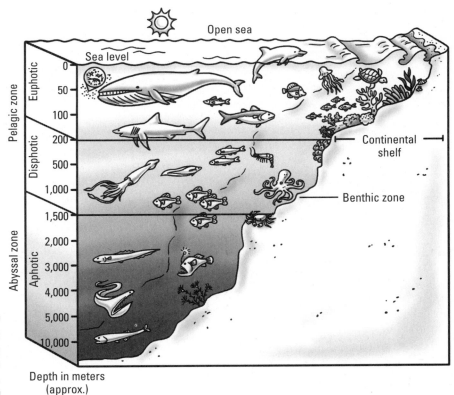

Figure 7-9:
The zones of the open sea.

Illustration by Lisa Reed

Chapter 8

Playing Well with Others: Population Biology

*T*he interactions that occur within an environment aren't restricted to molecules and energy. Living organisms within an ecosystem also interact with one another in complex ways: fighting to survive, sharing resources, and occasionally helping one another out, just to name a few.

Scientists who study the interactions that occur between living things are called *population biologists.* In this chapter, I explain the fundamental concepts of population biology that apply to environmental science. I start by describing the different characteristics of a group, or population, of organisms and then explain how population growth changes the way organisms interact with one another and with their environment. I also cover how ecosystems respond to and recover from disturbances and briefly introduce a few methods scientists use to study human populations.

Characterizing a Population

Scientists who study living organisms examine them from different perspectives of complexity. The simplest level is the *individual.* Each individual is a member of a population. Each *population* is made up of a group of individuals of the same species that occupy the same environment and interact with each other. Many different populations together make up a *community,* and many different communities interact with one another in an *ecosystem.* A group of ecosystems that interact with one another is called a *biome* (see Chapter 7), and all the biomes on the globe make up the Earth's *biosphere.*

Figure 8-1 helps you visualize the increasing complexity of living organisms on Earth, starting with an individual and going all the way up to the entire planet.

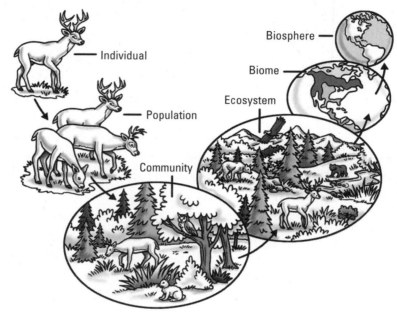

Illustration by Lisa Reed

Figure 8-1: Levels of complexity of living organisms on Earth.

Examining populations, specifically, is useful because they grow, decline, and respond to their environment together. Scientists use a few common measurements to characterize populations:

- **Size:** The *size* of a population is the number of individuals that make it up.

- **Density:** The *density* of a population is the number of individuals (population size) in relation to the area they inhabit.

- **Distribution:** The *distribution* of a population indicates where the individuals are located across the environment they occupy. For example, although 1,000 honeybees may live in your backyard, most of them stay in the hive, while only a few fly around to the flowers.

- **Sex ratio:** The *sex ratio* of a population is the number of males versus females (or vice versa).

- **Age structure:** The *age structure* of a population describes how many individuals fall into different age classes. For example, some populations consist mainly of young individuals, while others include individuals spread across many ages.

Predicting Population Change

Scientists who study living organisms recognize that populations change in response to different environmental and biological factors. One of the most obvious ways is through changes in population size. For example, some populations grow quickly, others fluctuate wildly, and still others maintain relatively steady numbers of individuals.

In this section, I describe some of the factors that affect population growth. I also explain the scientific models that population biologists use to characterize common trends in population growth and the reproductive strategies that link to these patterns.

Regulating populations with different growth factors

Scientists have identified many factors that control, or limit, the growth of populations in natural systems. For simplicity's sake, they divide these factors into two categories:

- **Density-dependent factors:** The influence of *density-dependent factors* on a population changes as the size of the population changes. For example, the amount of food available is a density-dependent factor. If the population grows larger, less food is available to support the individuals, so some of them don't survive. As individuals die, the population size decreases, leaving more food for survivors.

- **Nondensity-dependent factors:** *Nondensity-dependent factors* affect populations no matter how large or small the populations are. For example, natural disasters are a common nondensity-dependent factor. If a tornado or flood occurs and kills some individuals, it reduces the population size, regardless of how small or large the population was to begin with.

Modeling population growth

As I describe in Chapter 2, scientists often use scientific models to describe complex systems and interactions. In population studies, scientists use two models to describe how populations grow over time: the exponential growth model and the logistic growth model.

Two important concepts underlie both models of population growth:

✔ **Carrying capacity:** *Carrying capacity* is the number of individuals that the available resources of an environment can successfully support. In equations and models, the symbol K represents carrying capacity.

✔ **Limiting resource:** A *limiting resource* is a resource that organisms must have in order to survive and that is available only in limited quantity in their environment. Therefore, a limiting resource functions to limit population growth. Food and water are common limiting resources for animals.

In the following sections, I outline the two models of population growth that scientists use to predict changes in population size: the exponential growth model, which applies when there are no limiting resources affecting a population, and the logistic growth model, which applies when there are.

Exponential growth model

In the *exponential growth model,* population increase over time is a result of the number of individuals available to reproduce without regard to resource limits. In exponential growth, the population size increases at an exponential rate over time, continuing upward as shown in Figure 8-2.

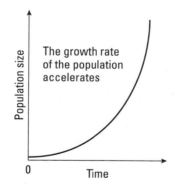

Figure 8-2:
Exponential
growth
model.

Illustration by Wiley, Composition Services Graphics

The line, or curve, you see in Figure 8-2 shows how quickly a population can grow when it doesn't face any limiting resources. The line creates a shape like the letter *J* and is sometimes called a *J-curve.*

Scientists also describe models with equations. The exponential growth model equation looks like this:

$$dN/dt = rN$$

The symbols in this equation represent concepts. Here's how to translate the equation into words: The change (d) in number of individuals (N) over a change (d) in time (t) equals the rate of increase (r) in number of individuals (N).

Logistic growth model

In reality, the growth of most populations depends at least in part on the available resources in their environments. To model more realistic population growth, scientists developed the *logistic growth model,* which illustrates how a population may increase exponentially until it reaches the carrying capacity of its environment. When a population's number reaches the carrying capacity, population growth slows down or stops altogether. Figure 8-3 illustrates the logistic growth model.

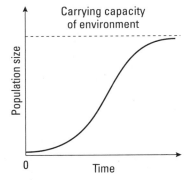

Figure 8-3:
Logistic
growth
model.

Illustration by Wiley, Composition Services Graphics

In the logistic growth model, population size levels off because the limiting resources restrain any further growth. This model applies in particular to populations that respond to density-dependent factors (refer to the earlier section "Regulating populations with different growth factors" for more on these factors). As you can see in the figure, the logistic growth model looks like the letter *S,* which is why it's often called an *S-curve.*

Scientists describe the logistic growth model with the following equation, which uses the same symbols as the exponential growth model (see the preceding section):

$$dN/dt = rN \left(1 - N/K\right)$$

This equation says that the change (d) in number of individuals (N) over a change (d) in time (t) equals the rate of increase (r) in number of individuals where population size (N) is a proportion of the carrying capacity (K). The best part about this equation is that it includes a way to factor in the negative feedback effect of a larger population relying on the same resources as a smaller population.

While the logistic growth model is often more descriptive of what occurs in reality than the exponential growth model, it still doesn't accurately describe what usually occurs in real life. What scientists have actually observed in nature is that populations seldom reach the carrying capacity and remain stable. Rather, they experience a pattern called *overshoot and die off*. As populations approach their carrying capacity, more offspring are born than the current resources can support; as a result, the population exceeds, or overshoots, the carrying capacity. When the population numbers exceed what the environment can support, some individuals suffer and die off because of the insufficient resources.

Figure 8-4 shows what the pattern of overshoot and die off looks like. A common situation that leads to this pattern is the variation in resource availability from year to year. For example, although plenty of food is available this spring while a population is reproducing, by the time the offspring are born, the food resources may have shifted enough (due to climate or seasonal changes) that they can't support all the new offspring.

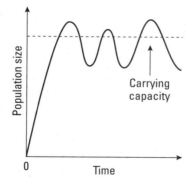

Figure 8-4:
Overshoot
and die off
pattern.

Illustration by Wiley, Composition Services Graphics

Reproducing strategies

In studying different species, scientists have realized that populations exhibit different *reproductive strategies* to increase or maintain population size.

Scientists classify some species as *K-selected species*. These species often keep their populations at or near carrying capacity (the *K* in *K-selected*). To maintain a fairly level population size, members of K-selected species usually have a long lifetime with few opportunities to reproduce and, thus, few offspring. These factors allow K-selected species to invest time and resources into the survival of their offspring.

Other species fall under the label of *r-selected species.* These species don't stay close to carrying capacity in terms of population size; instead, they grow rapidly (the *r* means "rate of increase" in population models) and exhibit extreme patterns of overshoot and die off (see the preceding section). Most r-selected species have short life spans and spend most of their lifetimes reproducing huge numbers of offspring, which they then leave to fend for themselves to survive.

Table 8-1 compares and contrasts the characteristics of K-selected and r-selected species; it also lists a few examples to give you an idea of what types of organisms I'm talking about here.

Table 8-1   Characteristics of K-Selected and r-Selected Species

Characteristic	*K-Selected Species*	*r-Selected Species*
Life span	Long	Short
Reproduction	Few chances to reproduce	Many chances to reproduce
Offspring	Few offspring; parents care for offspring	Many offspring; parents abandon offspring
Population growth	Slow, density-dependent	Rapid, nondensity-dependent
Population size	Stable, near carrying capacity	Variable, follows overshoot and die off pattern
Role in environment	Usually predators	Usually prey
Examples	Elephants, humans, most birds	Most insects, dandelions, mice

As with any model, the descriptions of K-selected and r-selected species represent the extreme ends of a wide spectrum of reproductive strategies. Many species exhibit characteristics of both K-selected strategies and r-selected strategies, so don't get bogged down by the characteristics I list in Table 8-1 when you're studying different species. Use them as guidelines, but be prepared to stray from them when necessary.

Watching Species Interact

Just like roommates, organisms that share an environment must find some way to share or divide up the available resources. After studying the many interactions between species in an ecosystem, scientists have come up with three categories for the types of interactions species have with one another: competition, predation, and symbiosis.

Table 8-2 provides a quick summary of the various ways that species interact with one another and specifies whether the interaction is helpful or harmful to the species involved. In the table, + indicates that the species benefits from the interaction, while – indicates that the species is harmed from it. The 0 indicates that the interaction doesn't harm or benefit the species. I cover the interactions listed in the table in more detail in the following sections.

Table 8-2	Species Interactions	
Interaction Type	*Species 1*	*Species 2*
Competition	–/+	–/+
Predation	+	–
Mutualism (a form of symbiosis)	+	+
Commensalism (a form of symbiosis)	+	0
Parasitism (a form of symbiosis)	+	–

Survival of the fittest: Competition

When two species need the same resource to survive, they have to compete over it. Competition over resources may occur among individuals of the same species — called *intraspecific competition.* Or it may occur between two different species — called *interspecific competition.*

Community ecologists who study the interaction of species with one another recognize that competition over a resource may define which species survive and which ones don't. The *competitive exclusion principle* states that two species competing for the same limiting resource can't coexist. Although the competitive exclusion principle makes sense in theory, in reality, separate species that depend on the same limiting resource have found ways to coexist. One way in which they do so is through a process called resource partitioning.

Resource partitioning is when two species divide a resource based on differences in behavior or lifestyle. For example, in some forests, certain species of owls and hawks both depend on the same food supply — small rodents. Owls hunt small rodents during the night, and hawks hunt small rodents during the day. Thanks to this difference in behavior, the two species can depend on and share the same limiting resources within an ecosystem.

Different species use different types of resource partitioning. The three main types are

- ✔ **Temporal:** In the example of the owl and the hawk, the species are using *temporal resource partitioning* because they use the same resource but at different times.

- ✔ **Morphological:** *Morphology* refers to the physical characteristics of an organism. Hence, *morphological resource partitioning* involves using similar but not the exact same resources thanks to differences in morphology. An example is when birds that depend on seeds and nuts for food have different sized and shaped beaks. These physical differences lead the birds with larger beaks to eat the larger seeds and nuts and the birds with smaller beaks to eat the smaller seeds and nuts.

- ✔ **Spatial:** Species sometimes use *spatial resource partitioning* within the same habitat. For example, because plants require the same limiting resources of water and nutrients, one species in a habitat may develop very deep roots to access water deep below ground, while another species keeps its roots close to the surface to access any rainfall or other surface water. In this way, the two plant species, while sharing a habitat, find different spaces within the habitat to access their needed resources.

Eat or be eaten: Predation

Predation is when one living organism uses another living organism as a food resource. Under this definition, scavengers, detrivores, and decomposers, all of which feed on already-dead organisms, aren't considered predators. Different species approach predation in different ways, though most species fall into the following broad categories:

- ✔ **True predators:** Animals that kill and consume another animal (called their *prey*) as food

- ✔ **Herbivores:** Animals that consume plants

- ✔ **Parasites:** Predators that consume part of another organism's tissues without killing the organism

Predator-prey interactions affect both population size and the distribution of populations across an ecosystem. Not to mention, they influence physical and behavioral characteristics of a species through natural selection (see Chapter 12).

In 1969, a scientist examining the food web of an intertidal ecosystem in Makah Bay, Washington, noticed that the ochre starfish was an important predator in maintaining the ecosystem's biodiversity and balance. When the starfish was removed, its most common prey, the mussel, outcompeted all the other species in its trophic level, and its population grew exponentially. Eventually, the mussels crowded out other species that lived in the intertidal ecosystem. The presence of the starfish predator leveled the playing field between mussels and their competitors. Without the starfish, the mussel population grew out of control, reducing its ecosystem's biodiversity.

From this study, scientists coined the term *keystone predators* for predators who play an important role in maintaining ecosystem stability. Keystone predators are one type of *keystone species*. They're named after the keystone of an archway — the stone at the center that keeps the entire arch from collapsing. In some ecosystems the keystone species is a plant (or other organism), not a predator.

Many species have evolved in response to *predation pressure,* or being preyed upon by predators. While some have become physically more able to escape predation (by running faster or climbing higher, for example), others have evolved more subtle ways to deter a would-be predator, including the following:

- ✔ **Chemical defenses:** Some plants and animals have developed chemical compounds that make them bad tasting or even poisonous when consumed. An example is the poison dart frog that secretes a poisonous chemical from its skin so that its predators become ill or even die after tasting it.

- ✔ **Mimicry:** Rather than develop their own defense, some species have evolved to look or act like a species that has good defenses. This defense mechanism is called *mimicry.* The species using mimicry benefits from looking like another species, one that has well-developed and well-known defenses (such as the toxins of the poison dart frog or the sting of a bee). Many different (harmless) butterfly species mimic their poisonous cousins in an attempt to scare away predators.

- ✔ **Camouflage:** Some organisms hide in plain sight by taking on the appearance of their immediate environment — a defense mechanism called *camouflage.* For example, stick insects have evolved to look just like a twig to help them hide from would-be predators.

In some cases, the predator-prey relationship becomes so important that it drives evolutionary change in both populations. *Coevolution* is the change in physical or behavioral characteristics of a species as a result of its interactions with another species over a long period of time (millions of years).

Scientists often point to the cheetah and the gazelle as examples of coevolution due to predator-prey interactions. Over time, the cheetah has developed a faster running speed to catch a gazelle. At the same time, the gazelle has become a little faster and (more importantly) better able to maintain high speeds for a very long distance, thus outrunning the cheetah with its endurance. (For more details on how predation and natural selection lead to evolutionary change, check out Chapter 12.)

Mutually surviving: Symbiosis

In some cases, not only do two species find a way to coexist or share resources, but one species depends on another to survive. This type of close interaction between species is called *symbiosis,* and it takes on the following different forms:

- **Mutualism:** *Mutualism* is a symbiotic relationship between two species in which both species benefit. For example, some plants experience mutualism with nitrogen-fixing bacteria that live on their roots. The plants provide food for the bacteria, and the bacteria fix nitrogen (see Chapter 6) for the plants to use.

- **Commensalism:** *Commensalism* is a symbiotic relationship between two species in which one benefits and the other neither benefits nor is harmed. An example of commensalism is when barnacles attach themselves to the skin of whales. The whale provides a place for the barnacles to live, while the barnacles have no effect on the whale.

- **Parasitism:** I describe *parasitism* in the previous section as a form of predation, but some scientists also consider it to be a form of symbiosis. A parasite benefits from consuming another organism's tissue, while the other organism (called the *host*) is harmed by the loss of nutrients. Parasites include insects that consume blood (ticks and mosquitoes, for example) and various worms that inhabit the intestines of humans and other animals.

From Indicator Species to Ecosystem Engineers: Filling Every Niche

Within any environment inhabited by multiple species, each organism must find a way to live from day to day. Different organisms survive best under different conditions. For example, some prefer more moisture or warmth; some live in treetops or underground. But they all have to find food to eat, water to drink, and a safe place to sleep. How a population uses these resources and affects its environment and other species in the ecosystem is called its *niche*.

Every species has an ideal range of environmental conditions, specifically temperature and precipitation ranges, under which it thrives. These *tolerance limits* of temperature and precipitation define the *fundamental niche* of that species; in other words, the fundamental niche is all the places it could successfully inhabit if there were no other species to compete with. Many species can survive at the edges of their fundamental niche, but they prefer to live under their ideal conditions.

Of course, ideal conditions are rare, and many other organisms have to share the environment, so many species end up living in what scientists call a *realized niche,* which is the smaller portion of the environment that a species inhabits within its tolerance limits.

Species with a very narrow range of tolerance limits often act as indicator species. *Indicator species* are species that are very sensitive to changes in their environment. When the environmental factors change, even slightly, such as when lake water becomes a little warmer than usual, the indicator species suffers. A change in the population of an indicator species signals to scientists that something has changed or is out of balance in their ecosystem.

To make the best use of the wide range of conditions in an ecosystem, some organisms become *niche generalists* and others become *niche specialists:*

- ✔ **Niche generalists:** These organisms adapt to live in a variety of habitats.

- ✔ **Niche specialists:** These organisms can successfully survive only in a very confined set of conditions. As a result, they're more vulnerable when environmental conditions change, but as long as conditions are stable, they thrive.

Scientists sometimes use the term *niche* to describe the lifestyle of an organism and the role it plays in its community (I provide more details on communities in the next section). For example, being a keystone predator is a particular ecological niche. Other important ecological niches include

- ✔ **Pioneer species:** Certain plant species can tolerate environments with a low supply of soil nutrients. These species use this tolerance to populate areas where nothing else can grow due to the absence of nutrients. Because they're usually the first organisms to populate an environment, scientists call them *pioneer species.* After the pioneer species have established themselves in an environment, other plants and animals follow, being drawn to the increased cycling of nutrients and organic matter into a previously sterile location.

- ✔ **Ecosystem engineers:** Species that change the environment in a way that provides a new or improved habitat for other species are called *ecosystem engineers.* Beavers are common examples of ecosystem engineers. By building dams, beavers create new habitats in and around the river that other organisms benefit from.

As organisms from different species fill the various niches of an ecosystem, they create a community with unique structural characteristics, which I discuss in the next section.

Working Together: Biological Communities

Scientists who study communities look at the number of organisms, the number of species, and the distribution of the plants and animals across the environments that they inhabit. Two measures that help scientists study communities and observe how they change over time or across their environment are abundance and richness:

- ✔ **Abundance:** The total number of organisms in a community
- ✔ **Richness:** The number of different species that make up a community

Although these two concepts may seem similar, they measure two different things. For example, an anthill has a high abundance (many individuals) but low richness (only one species). Conversely, a tropical fish tank may have only a few individuals (low abundance), but it may have many different species (high richness).

The *ecological structure* of a community describes where the organisms are physically located across their environment and in relation to one another. For example, a community's ecological structure may be random, such as wildflowers scattered across a field. Or it may be clustered, such as a school of fish swimming together through the ocean. Some communities, such as communities of prairie dogs, even exhibit an orderly or uniform structure in which the individuals spread out at even intervals across the landscape.

Succeeding through different stages

Communities of organisms change over time in terms of how they're structured and what species they contain. In particular, communities shift and change over time in a process called *succession*. While many types of communities experience change through succession, the concept is most easily illustrated with a plant community. The two types of plant succession are

- ✔ **Primary succession:** Pioneer species move into an environment that has no soil or organic matter present. After they have established themselves in their new environment, they begin to cycle nutrients and organic matter, creating soil and providing a healthy base for other (non-pioneer) species to follow. After enough organic matter is present in the soil to support them, tree seedlings take hold and grow, taking up space and blocking the sun from the smaller plants. Figure 8-5 illustrates what primary succession looks like over time.

✔ **Secondary succession:** This type of succession occurs when soil is present and plants have been growing but are removed (usually by a disturbance of some kind, which I describe in the next section). Similar to primary succession, pioneer species are the first to populate the cleared area, followed quickly by trees and larger plants.

Figure 8-5:
Stages of
succession
in plant
communities.

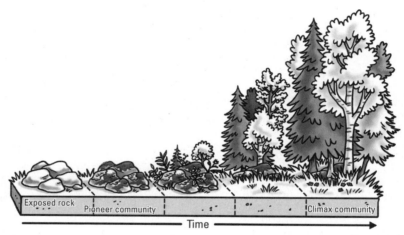

Exposed rock   Pioneer community   Climax community

Time

Illustration by Lisa Reed

The end point of succession for plant communities, whether it's primary or secondary, is the *climax community*, or the mature phase that plant communities reach when they're left alone to grow and change.

Responding to disturbances

Of course, no community is left alone for very long! Even without humans or other animals, plant communities often have to deal with unexpected changes, such as floods, fires, and windstorms. These events that disrupt the community are called *disturbances*.

Disturbances can change the structure of plant communities in several different ways, including knocking down trees or removing plants and animals from large parts of an environment. Although your instinct may tell you to protect communities from such disruptive events, scientists have found that some communities actually prefer to be disturbed every now and then!

CASE STUDY

Unpredictable response: Rebuilding after a volcanic eruption

In 1980, Mount Saint Helens, a volcano in the northwest corner of the U.S., erupted explosively. The heat, gas, and ash that spewed from the mountaintop destroyed 230 square miles of the landscape. What had previously been dense, richly populated mountain forests, prairies, lakes, and streams was left looking like the surface of the moon, covered by an ash layer 5 or more inches thick.

Biologists and community ecologists studying the ecosystems surrounding Mount Saint Helens expected that the community surrounding the volcano would take many years to proceed through the predicted stages of succession before it resembled its old self again. According to rules of community succession, pioneer species would come in first, paving the way for herbaceous (weed-like) plants and eventually creating an ecosystem that could support insects, mammals, and other organisms again.

Imagine their surprise when they observed a thriving ecosystem at Mount Saint Helens only a year later! The pioneer species of plants weren't the lichens that scientists had expected but rather the nitrogen-fixing lupines, which worked to prepare the soil for other types of vegetation. The presence of the lupines provided a food source for some insects, whose populations exploded without the presence of their predators.

New and interesting species relationships also developed. Pocket gophers who lived underground began digging tunnels through the ash and soil layers below, bringing organic matter to the surface and spreading important nutrients and nitrogen-fixing bacteria to support plant growth. The gopher dens also became shelters for small amphibians, such as salamanders, frogs, and toads. These creatures couldn't survive long on the hot dry surface, but they could use the gopher tunnels for shelter and for getting around the ecosystem.

These and other examples illustrated to community ecologists that living organisms and ecosystems are highly adaptable in the face of disturbance and may not always respond in predictable ways. In fact, communities often respond unpredictably with a single goal in mind: survival.

Source: PBS *NOVA,* "Mount Saint Helens: Back from the Dead"

CASE STUDY

Take for example the ponderosa pine tree forests in the western U.S. Under natural conditions, these forests experience common (every five years or so) relatively small fires that burn away the low-growing plants and grasses but only gently burn the bark of the trees, leaving them to recover and continue growing. The U.S. Forest Service's efforts to reduce forest fires interrupted this natural cycle of disturbance. As a result, instead of an occasional small fire, when fires did occur in these forests, they grew to a much larger size and caused a lot more damage — destroying every tree in their path. The buildup of low-growing plants and small trees gave fuel to the fires, making them strong enough to destroy pine trees that had weathered smaller fires for more than 200 years.

Scientists have since realized that fire is a welcome occasional disturbance in these communities. In fact, some species are considered *disturbance-adapted species* because they have adapted and evolved to depend on the occasional community disturbance. In the case of the lodgepole pine, for example, the trees need fire in order to open their pine cones and release seeds.

When any type of disturbance disrupts a community, the organisms in the community must respond quickly and effectively if they want to survive. This ability to respond quickly (in other words, to bounce back) from a disturbance is called *resilience*.

The resilience of a specific community often depends on the complexity and structure of that community. The more diverse a community is, the more easily it can respond quickly to a disturbance. For example, a community with a high degree of diversity has many species available to refill the ecological niches left empty after a disturbance.

The People Principle: Human Population Biology

Although human beings are organisms, they don't always respond to environmental conditions as predictably as other organisms do. After all, various cultural factors influence how human populations interact with their environment. Hence, the study of human populations, called *demography,* reaches beyond population biology to include culture, technology, and other factors that influence human populations.

In this section, I describe some measurements and other methods that scientists use to study human populations. (In case you're wondering, scientists who study human populations are called *demographers.*)

Tracking exponential growth

As with other organisms, human population size is the result of how many individuals are born, how many survive, and how many die. Until recently, human population growth, like that of most other species, was closely linked to the ecosystem individual humans inhabited. Population growth was slow and fairly steady in direct response to environmental limits such as the carrying capacity. However, the technological innovation and advancement of the human species changed this relationship. Figure 8-6 illustrates the general trend in human population growth over the last few thousand years.

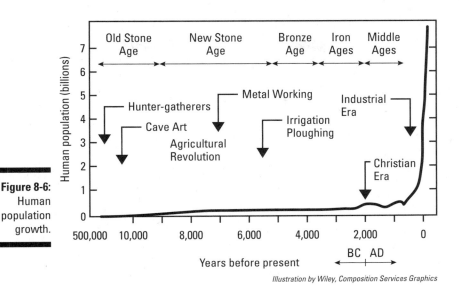

Figure 8-6:
Human
population
growth.

Illustration by Wiley, Composition Services Graphics

The modern, industrial age has allowed the total population to grow exponentially for the first time in human history. The environment no longer limits human populations the way it does for most species. What environmental scientists see now is that human populations are changing and shaping their environment to meet their needs — the needs of a growing population.

Calculating population size

As scientists study the effects of an increasing human population on the planet's resources, they realize that the overall exponential growth illustrated in Figure 8-6 isn't occurring in every society. Some human populations are shrinking, others are growing, and still others remain stable (neither shrinking nor growing).

Calculating accurate population size for different nations or regions is difficult to do since people are constantly being born and dying, as well as moving around. Even so, demographers use a few common measurements to help them estimate human population size:

✔ **Crude birth rate:** The number of babies born in a year per 1,000 people

✔ **Crude death rate:** The number of deaths per 1,000 people in a year; sometimes called the *crude mortality rate*

> ✔ **Total fertility rate:** The estimated number of children born to an average woman in that population over her entire lifetime
>
> ✔ **Doubling time:** An estimate of the number of years a population needs to double its current size

By subtracting the crude death rate from the crude birth rate, you can calculate the *natural increase of a population,* but that particular calculation doesn't account for either immigration or emigration. *Emigration* (moving out of a country) and *immigration* (moving into a country) play a much larger role in studies of human populations than in studies of other animals. Humans are highly mobile and move around not just within their immediate environment but often into completely new environments. To include these important factors, demographers calculate the *total growth rate* with this equation:

Total growth rate = (Crude births + Immigration) – (Crude deaths + Emigration)

When a country has *zero population growth,* the number of births and immigrants equals the number of deaths and emigrants. To maintain zero population growth, a population needs only *replacement level fertility,* which is usually about two children per childbearing woman — enough new people to replace their parents when they die, therefore keeping the total population stable.

Diagramming age structure

Population size isn't the only informative way to measure human populations. Demographers also look at the distribution of people across different ages within a population. The *age structure,* or distribution of individuals across five-year age ranges, can indicate whether a population is likely to grow, stabilize, or decline in the near future.

When demographers diagram age structure data, they do so using a *population pyramid.* Contrary to the name, however, only some population pyramids are actually shaped like pyramids, such as the one illustrated in Figure 8-7a. In this pyramid, more individuals are in younger age groups than in older ones. From this age structure, you can estimate that the population is going to increase as the large number of young people start to have children of their own.

Eventually, a growing population may have an age structure more like the one shown in Figure 8-7b, in which the individuals are more evenly distributed across the age ranges. In such a population, total population growth is slow, or approaching zero and stable, because the number of deaths is well-balanced by the number of births.

Countries that have shrinking population numbers have population pyramids that look like the one shown in Figure 8-7c, in which the base of the diagram is narrower than the middle or top. From this age structure diagram, you can infer that the population has fewer young people and a low fertility rate. If the fertility rate is below replacement level, then the population will become smaller and smaller over time.

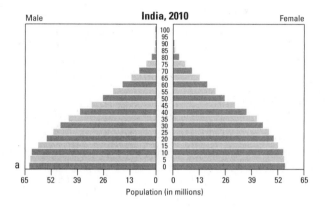

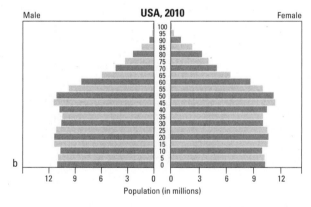

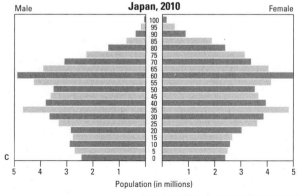

Figure 8-7:
Age
structure
diagrams for
India (a), the
U.S. (b), and
Japan (c).

Illustration by Wiley, Composition Services Graphics

Experiencing a demographic transition

Demographers have recognized a pattern in population growth as nations move into more developed, or industrialized, ways of life. The *theory of demographic transition* states that as the economy of a country moves from being mostly subsistence based (as in hunting or farming) to industrial, the population will increase and stabilize. The *demographic transition model,* shown in Figure 8-8, is a visual representation of the theory of demographic transition. It's based on observations of European countries before, during, and after the Industrial Revolution in the late 1800s and early 1900s.

Figure 8-8 illustrates the relationship between births, deaths, and total population for each stage of the demographic transition that I describe here:

- ✔ **Stage 1 (Pre-modern):** In Stage 1, population growth is slow or steady with a high birth rate and a high death rate. The total population size is relatively small, and life expectancy is short. The subsistence-based economy (such as hunting, fishing, or farming) is closely tied to the available resources of the immediate environment.

- ✔ **Stage 2 (Industrializing):** In Stage 2, population growth increases exponentially as the birth rate stays high and the death rate declines. The death rate decreases because of increased sanitation and technological advances in healthcare and medicine during the initial phase of industrialization. Also, overall wealth increases with access to a wider variety of resources and products.

- ✔ **Stage 3 (Industrial):** After the initial surge in population numbers during Stage 2, Stage 3 shows a slowing down in population growth as the birth rate drops closer to the death rate. At this stage, the population growth is more stable though at higher total numbers than in Stage 1. Scientists attribute the lower birth rate to the financial burden of supporting more children in an industrial society, the availability of birth control, and the increased amount of time spent on education.

- ✔ **Stage 4 (Post-industrial):** The fourth and final stage of the demographic transition has low birth and death rates. Population growth remains stable, but it shows a higher number of individuals than in previous stages. Scientists are still adjusting this stage, where many currently industrialized nations fall in the model, based on new data they collect. So far, they've observed that some nations in Stage 4 are actually beginning to experience population decline.

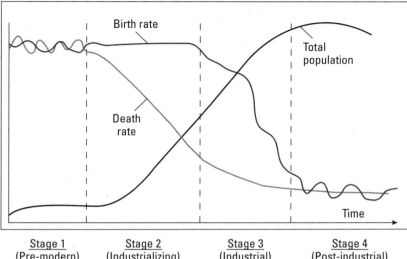

The Demographic Transition Model

Figure 8-8:
The four
stages
of demo-
graphic
transition.

Stage 1	Stage 2	Stage 3	Stage 4
(Pre-modern)	(Industrializing)	(Industrial)	(Post-industrial)
High birth	High birth	Declining birth	Equal birth and
and death	rates but	and death rates	death rates means
rates	declining	but population	population is stable
	death rates	still growing	

Illustration by Wiley, Composition Services Graphics

Getting caught in a demographic trap

In theory, the demographic transition model describes a smooth path from pre-modern to industrial characteristics of population growth. In reality, however, things don't always go so smoothly. For instance, many of the current developing nations are caught in the *demographic trap;* that is, they're stuck in Stage 2 of the demographic transition model (see the preceding section).

The result of the demographic trap is that population numbers continue to grow as the death rate drops and the birth rate remains high, but the population doesn't move smoothly into Stage 3, where the birth rate also drops. As a result, the population experiences explosive growth beyond what the developing nation's economy and infrastructure can comfortably support. In turn, many people in these nations remain impoverished, even as technological advances and economic development provide access to more wealth and resources.

One suggestion for helping countries complete the demographic transition has to do with social justice, which I describe in detail in Chapter 20.

Understanding human impacts

Environmental scientists are especially concerned with the effect that increasing population numbers have on the Earth's ecology. In Chapter 5, I introduce the concept of the ecological footprint, which is one tool environmental scientists use to estimate how each person impacts the Earth's resources, but here I focus on the impact whole populations have on the Earth's resources. To calculate the effect of an entire population on its environment, you can use the *IPAT formula*. The IPAT formula looks like this:

Impact (I) = Population size (P) + Affluence (A) + Technology

This equation is conceptual rather than mathematical, which means it encompasses ideas about what impacts the environment — and some of these ideas are difficult to quantify with numbers. To better understand how you can use this formula, try looking at each component separately:

- **Population size:** Population size is the easiest component of the IPAT formula to quantify. After all, you can safely assume that two people impact the environment twice as much as one person.

- **Affluence:** *Affluence* in the IPAT formula refers to the consumption of goods and services. Affluence varies greatly from one society to another. For example, an upper-class individual in an industrialized nation with greater affluence likely has a greater impact on the environment than two or three combined individuals in a developing nation.

- **Technology:** Technology in the IPAT formula represents how many resources humans use to create the affluence of their society and how much they damage or degrade the environment in the process. Because newer technologies often seek to improve the environment or reduce environmental degradation, some scientists have begun to use the term *destructive technology* rather than just *technology* in the IPAT formula.

The IPAT formula explains how countries or communities with greater wealth and technology have a larger impact on their environment than those that are poor. Increasing technology that provides increasing wealth often leads to an increasing need for energy and other resources. These increases, in turn, lead to problems of resource distribution and environmental quality (see Parts III and IV).

Part III

Getting Your Needs Met: Earth's Natural Resources

The 5th Wave   By Rich Tennant

"Take it easy everyone. Let's just hope the wind currents carry this thing out of here."

In this part . . .

Human beings everywhere depend on the environment to take care of their basic needs: clean water, fresh air, food, and warmth. In this part, you see how the Earth provides resources to meet these needs and how environmental scientists work to find ways to make Earth's natural resources last as long as possible. After all, billions of people have to share the resources of one Earth for many years to come.

This part describes water, land, and food resources and introduces the idea of biodiversity as a valuable resource. Of course, no discussion of resources is complete without considering energy sources, so you also find information about the limited fossil fuel resources and the alternative energy sources that are available for the future.

Chapter 9

Water, Water Everywhere: Freshwater Resources

*W*ater is quite possibly the most important resource the environment provides. All living things need water in order to survive. The good news is that the Earth contains a lot of water. The problem is that more than 97 percent of the Earth's water is salt water in the oceans and, therefore, isn't drinkable. Living things need *fresh water,* water without high levels of salt, to live.

In this chapter, I explain where freshwater resources are located and how humans access the fresh water they need to drink, bathe, and grow food. I also describe regions of the world that don't have access to adequate supplies of fresh water and explain how environmental scientists seek to create fresh water from salt water.

Thirsty for More: The Never-Ending Need for Fresh Water

Environmental scientists approach freshwater resources with an understanding that both human societies and surrounding ecosystems need access to a certain amount of water. Thus, people need to share the available fresh water with each other and with all the other organisms that inhabit the environment. In this section, I explain how scientists define the use of fresh water in the environment and describe the most common ways in which people use freshwater resources.

Withdrawal versus consumption

Scientists define the use of the Earth's freshwater resources in two ways:

- ✔ **Withdrawal:** Water *withdrawal* measures the total amount of water removed from its natural source (such as a lake or river). Water that's withdrawn can be used and returned to its source for reuse.

- ✔ **Consumption:** Water *consumption* measures the amount of water lost (through evaporation, absorption, or chemical transformation) during use. Water that's consumed can't be returned to its source and reused.

Water that's withdrawn but not consumed may be degraded or polluted. When this water is returned to its natural source, it's no longer suitable for human or ecosystem use, but it hasn't been consumed. Chapter 16 covers the environmental issues related to water pollution.

Scarcity and stress

In many cases, fresh water is a *renewable resource*, meaning that it can be recycled and reused repeatedly or that its supplies will be naturally replenished after people (and other organisms) use it. But sometimes the need for water exceeds the availability of local fresh water. This situation — where there isn't enough water to meet the needs of people and ecosystems — is called *water scarcity*. Water scarcity can occur for two reasons:

- ✔ Not enough local water is available to meet the needs of people and ecosystems.

- ✔ The available water is polluted or otherwise can't be used to meet every need.

Situations of water scarcity can lead to *water stress*, which is when inadequate water supply leads to competition and conflict as people try to find ways to meet their water needs. Water stress is most common in regions where the amount of fresh water per person is low, and it can increase even more during years of *drought* (when seasonal water from rain is absent or lower than expected).

Meeting human water needs

Scientists divide the different ways people use water into three categories: agricultural, domestic (or household), and industrial. I describe these three uses in the following sections.

Watering the crops: Agricultural uses

If you've ever owned a houseplant or tried to maintain a green lawn or garden, you know that plants need water. Thus, you may not be surprised to hear that *agriculture,* the growing of plants as food, is the largest consumer of fresh water on Earth, accounting for nearly 70 percent of all freshwater withdrawal. (As I describe in Chapter 11, farming food requires both fertile land and water.)

One of the biggest challenges of farming in some regions of the world is locating enough fresh water to support crops. In the drive to meet the food needs of growing human populations, farmers have extended their croplands into drier regions that are farther from natural, seasonal sources of water. As a result, farmers have to build *irrigation systems* to bring water to the crops in these drier regions.

Irrigation systems come in many different forms, depending on the landscape, the regional water availability, and the water needs of the crops. A few of the most common types of irrigation systems are

✔ **Furrow irrigation:** *Furrow irrigation* involves digging *furrows,* or channels, alongside rows of crops. It's one of the oldest methods of irrigation and was used by ancient civilizations in Egypt and Mesopotamia. By digging shallow ditches along a gentle slope, farmers rely on the pull of gravity to transport the water from a nearby river or stream into their crop fields. The main problem with furrow irrigation is that it isn't the most efficient way to water crops. In some regions, as much as 35 percent of the water transported to the crops evaporates or runs off the field without being absorbed into the soil.

✔ **Flood irrigation:** *Flood irrigation* uses a natural source of nearby flowing water, such as a river or stream, and periodically diverts the water to flood agricultural fields. This irrigation method allows the water to completely cover and soak into the fields. It's more efficient than using furrows because it loses only 15 to 20 percent of the water to evaporation or runoff.

✔ **Drip irrigation:** *Drip irrigation* applies small amounts of water more directly to the plants that need it. This localized irrigation system uses hoses and pipes to drip water onto (or just below) the soil surface. Losing only 5 percent of the water to evaporation, drip systems are very efficient. They work best in fields that don't need to be plowed every season because the drip hoses are woven through the field at or below the soil surface.

✔ **Sprinklers:** Like drip irrigation systems, *sprinklers* use pipes and hoses to move water. But unlike drip irrigation systems, sprinklers spray water over the fields from above and, thus, require a form of energy to pump the water through the pipes. The efficiency of sprinkler systems varies: Some systems lose up to 25 percent of the water, while others lose only about 5 percent. In large agricultural fields, farmers often use sprinklers that are mounted on wheeled systems that move through the fields. Another common sprinkler is the *traveling sprinkler system,* which sprays water from a long arm that pivots around a center point. If you've ever seen a bird's-eye view of agricultural fields, you may have noticed fields laid out in circles across the landscape; this circular layout is a result of the traveling sprinkler system, which effectively waters a circle of crops from its center pivot point.

Determining which irrigation system is most sustainable for a particular region depends on many factors, including the availability of water and energy resources, the size and layout of the coverage area, the system costs, and the overall efficiency (which depends, in part, on local soil and weather conditions).

The development of *hydroponic agriculture* offers a new approach to reducing agricultural water use. A hydroponic system grows crops in a greenhouse without using soil. Instead of soil, the crops are "planted" in nutrient-rich water. The water not used by the plants is recycled and reused, and the growing conditions are controlled from above (by the greenhouse) and below (by the nutrient solution) to be ideal for maximum crop production. Hydroponic agriculture requires extra costs upfront to set up the greenhouse facility, but in the long term, the method saves water and soil resources and also reduces the need for pesticides.

Washing and flushing: Domestic uses

The second largest consumer of fresh water is you (and every other person in the U.S.). Every day you drink water, brush your teeth, wash your clothes, flush the toilet, and bathe. These types of household or *domestic* water use account for more than 10 percent of the freshwater use in the U.S. Figure 9-1 illustrates where the average U.S. household consumes fresh water.

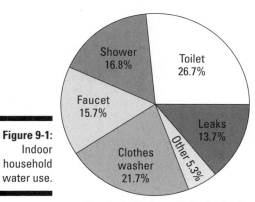

Figure 9-1:
Indoor
household
water use.

The percentages in Figure 9-1 are averages for the U.S. The actual use of water in a household depends on what type of plumbing and sewage infrastructure it has. For example, in regions that don't have indoor plumbing, households don't use water for flushing toilets. In addition, the graph in Figure 9-1 doesn't include outdoor household water use, such as watering the lawn or garden, which accounts for about 25 percent of total household use (on average). I explain how to reduce the consumption of household water in the later section "Conserving Fresh Water" and in the sidebar "Do-it-yourself water conservation."

Keeping things cool: Industrial uses

Various industries use water to produce energy, refine metal, and manufacture products. Most of the industrial water use in the U.S. is for the production of electricity, either through hydropower dams or at power plants. Although the capture of energy from moving water, called *hydropower,* doesn't consume water because the water is still available for other uses in the ecosystem, other sources of electricity do consume water. (For more details on hydropower, check out Chapter 14.)

Nuclear reactors and coal power plants, for example, consume fresh water, meaning that the water these industries use for the production of electricity is no longer available for other uses. Both nuclear and coal plants transform water into steam to power engines that generate electricity. In the process, most of the steam is lost to the atmosphere; only some of it is collected, converted back to a liquid, and returned to its source.

Other industrial uses result in *water waste,* or the pollution of water by metals or chemicals. Mining for ores (see Chapter 13) requires a large amount of water to rinse unwanted minerals away from the desired metal resource. Once mined, these metal resources must be refined and manufactured into products, such as aluminum foil, appliances, and cars. This type of industrial water use is a common source of water pollution, which I describe in Chapter 16.

Finding Fresh Water

Through the hydrologic cycle (see Chapter 6), water constantly moves among the oceans, the atmosphere, the Earth's surface, and underground. Of all this water, only 3 percent is fresh water. Most of this fresh water (about 69 percent of it) is currently stored as ice in glaciers and ice sheets; the rest is stored and flowing as lakes, ponds, and rivers (about 0.3 percent) or as groundwater beneath the Earth's surface (about 30 percent). Less than 1 percent of the world's fresh water is located in the atmosphere (in the form of precipitation).

To effectively use these freshwater resources, people must find ways to control the water flowing on the Earth's surface or access the groundwater below. In this section, I describe the various ways to access the Earth's fresh water.

Diverting surface flow

To use the fresh water that flows along Earth's surface as rivers and streams, people change where it flows, or *divert* it. Diversion projects are basically just manmade structures that take water from one area and bring it to areas that need it. Two of the most common diversion projects are

- **Aqueducts:** *Aqueducts* are canals or pipelines that carry water from its natural source to an area that needs it. Both New York City and Los Angeles use aqueducts to divert fresh water from distant sources (the Catskill Mountains for New York and the Colorado River for Los Angeles). This type of water diversion goes all the way back to ancient Greece and the Roman Empire, though today's versions are much more efficient.

  Although diverting fresh water through aqueducts solves the problem of supplying water to large urban areas, doing so also creates problems. For one, the diverted water is no longer available in the ecosystem it originally flowed through for the organisms who depend on that water source for survival. Using aqueducts is also likely to negatively affect people outside the urban center who depend on the natural flow of the water for their fresh water.

✔ **Dams:** A *dam* is a structure that blocks the flow of a river, creating a large *reservoir,* or lake behind it where the water is stored for human use. Humans use dams for many purposes, including the production of electricity through hydropower, which I describe in Chapter 14.

But as with any manmade change to a natural system, creating dams has some negative consequences. For example, because of the way they're constructed, dams flood large areas of land behind them, upriver, where the reservoir is located. In some instances, this flooding destroys villages and important ecosystems. Dams also obstruct the natural flow of water and sediment downstream, and this obstruction, in turn, affects fish migration and changes the natural evolution of river habitats.

Tapping what flows below: Groundwater

Most of the fresh water that people access flows underground. The fresh water that flows through rocks and open spaces below the Earth's surface is called *groundwater.* Although the ground you walk on is solid, spaces between the particles of sediment, or even within certain types of rock, allow water to move from the surface into underground storage spaces called *aquifers.* As you can see in Figure 9-2, the two types of aquifers are

✔ **Unconfined aquifers:** Water in an *unconfined aquifer* is stored in *permeable* rocks and sediment through which it can flow freely. Hence, water in this type of aquifer can flow to plant roots or bubble up to the surface as a spring.

The *water table* is the boundary between the water-filled rock and sediment of an aquifer and the dry rock and sediment above it. Water that seeps into the ground through the water table when it rains refills, or *recharges,* the groundwater in unconfined aquifers.

✔ **Confined aquifers:** *Confined aquifers* are surrounded by *impermeable* layers of rock that don't allow water to move through them. Thus, confined aquifers create underground storage containers for the water they contain. Because impermeable rock layers surround confined aquifers, they have a specific *area of recharge,* where fresh water from rainfall can enter and refill the aquifer.

To withdraw groundwater stored in both types of aquifers, people dig *wells.* Unfortunately, the rate of recharge for most groundwater aquifers is much slower than the rate of withdrawal through wells to meet human water needs. As a result, many existing wells are now *dry wells,* where no more water can be drawn, and cones of depression form in the water table. A *cone of depression* is an area where the water table dips because water has been withdrawn from that area of the aquifer faster than it could be recharged (see Figure 9-3).

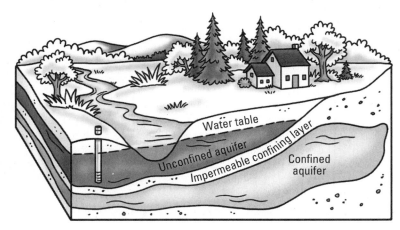

Figure 9-2:
The two types of aquifers.

Illustration by Lisa Reed

Withdrawing groundwater from aquifers faster than it can be recharged can result in *saltwater intrusions* in coastal regions, where fresh water underground contacts the salt water of the ocean nearby. When a cone of depression occurs, the space created by the withdrawal of fresh water may fill up with salt water rather than fresh groundwater, hence the name *saltwater intrusion*. Figure 9-3 shows both a cone of depression and a saltwater intrusion.

After salt water has intruded into an aquifer, the aquifer is no longer a source of fresh water for the people and ecosystems that depend on it.

Figure 9-3:
A fresh groundwater source with a cone of depression and saltwater intrusion.

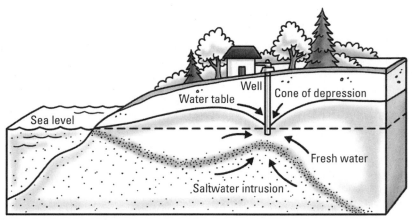

Illustration by Lisa Reed

Conserving Fresh Water

One way to meet the freshwater needs of people and ecosystems is to use techniques of water conservation. *Water conservation* is the process of using less water to begin with and recycling or reusing as much water as possible. The goal of water conservation is to maintain a freshwater supply that can meet the needs of as many people as possible for as long as possible.

Technological innovation has helped achieve much of the water conservation happening today. For example, water-efficient showerheads and toilets reduce the amount of household water use in many homes. A dual-flush toilet offers the user an option between a normal flush (approximately 2 gallons of water) for solid waste and a lighter flush (about 1 gallon) for liquid waste. Manufacturers are also producing more water-efficient washing machines and dishwashers.

Do-it-yourself water conservation

Conserving water is one of the easiest ways to reduce your impact on local water resources and the other organisms that depend on them. Here are some ways you can start conserving fresh water today:

✔ Turn off the faucet while brushing your teeth or shaving.

✔ Wash only full loads of laundry.

✔ Position sprinklers to water the lawn and garden, not the sidewalk or driveway.

✔ Plant native shrubs and groundcovers rather than grass in your landscaping.

✔ Allow your lawn to go dormant for a few months in the summer.

✔ Compost food waste instead of using the garbage disposal.

✔ Repair leaky faucets indoors and outdoors.

✔ Install aerators on all your faucets.

✔ Upgrade to more-water-efficient appliances, including toilets, showerheads, washing machines, refrigerators, and dishwashers.

✔ Collect rainwater from your roof in rain barrels and reuse it to water your garden.

✔ Rinse vegetables in a dish of water and then dump that water in your houseplant or garden.

Another approach to water conservation is to recycle fresh water within your home through a *greywater reuse system.* The term *greywater* refers to the wastewater from your sinks, showers, and washing machines (everything except your toilet water, which is considered sewage). Although you can't use greywater for drinking, you can use it to water your lawn or flush your toilet. A greywater reuse system filters your home's greywater so that it can be reused for other domestic freshwater needs.

Creating Fresh Water

In some regions of the world, even practicing water conservation can't meet all the freshwater needs of people and ecosystems. The Middle East, in particular, faces extreme water scarcity and water stress. One approach to dealing with these issues is *desalinization,* the process of removing salt from salt water to create fresh water. Scientists have developed two ways to achieve desalinization:

- ✔ **Distillation:** Scientists heat the salt water enough that the water molecules evaporate, leaving the salt behind. The scientists then cool the evaporated water, or steam, so that it condenses into a salt-free liquid. Figure 9-4 illustrates this process.

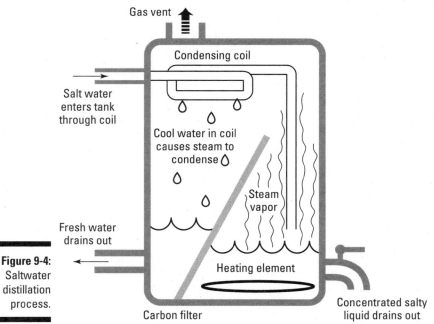

Gas vent

Condensing coil

Salt water enters tank through coil

Cool water in coil causes steam to condense

Steam vapor

Fresh water drains out

Heating element

Figure 9-4: Saltwater distillation process.

Carbon filter

Concentrated salty liquid drains out

Illustration by Wiley, Composition Services Graphics

✔ **Reverse osmosis:** Scientists expose the salt water to increasing pressure, which forces the water to move through a membrane. The membrane allows water molecules to move through it while blocking the salts. Like a filter, the membrane removes the salt and produces fresh water. Figure 9-5 shows what this process looks like.

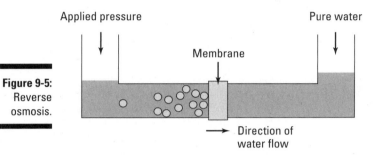

Figure 9-5:
Reverse
osmosis.

Illustration by Wiley, Composition Sewrvices Graphics

The growing human population will need increasingly more fresh water to meet its needs. Ensuring that every person and ecosystem have as much fresh water as they need will require a combination of water conservation and desalinization technologies.

Chapter 10

This Land Belongs to You and Me: Land as a Resource

M any of the resources you depend on, including raw materials, food, and fuel resources, come from the land. But the land itself is also a resource — for the raw materials it provides, the ecosystems it supports, and the space it offers human societies to expand their roads, towns, and cities.

In this chapter, I describe how people deal with shared land (that is, land that doesn't belong to any one person), including how they use it for recreation and resource harvesting. I also explain how increasing urban and suburban development endangers this limited resource and describe methods of smart growth that seek to conserve land resources.

Sharing Land

You may own the land on which your house is built, but most of the land space you inhabit is *shared land* — land that many different people have a right to access. Environmental scientists have recognized that if no one owns the land, then no one takes on the responsibility of caring for and managing its resources. As a result, human populations quickly deplete those resources without thinking twice.

In this section, I explain how sharing land can lead to resource depletion, and I describe the different ways in which people classify and manage shared land around the world.

Watching a tragedy unfold: Land resource depletion

The *tragedy of the commons* scenario describes how all the people who have access to common areas or shared public space will use it to meet their current needs, thus allowing the land to be damaged and the resource depleted. The assumption in this scenario is that every person who uses a shared resource acts primarily in self-interest, seeking short-term gain without planning and managing the resource to sustain it for the future.

The tragedy is that this scenario of resource destruction is avoidable, if only the people sharing the resource worked together and practiced sustainable management methods. In Chapter 23, I list ten examples of the tragedy of the commons scenario in action, including the near extinction of bluefin tuna and the irreparable destruction of vast expanses of rainforests.

These and other tragedy of the commons scenarios illustrate that publicly shared land must be regulated in such a way that its resources can be managed sustainably — providing support for the people and other organisms in the ecosystem now and into the future. Acknowledging that unregulated land sharing is a problem has led to today's methods for public land classification and management (see the next section for details).

Classifying shared land

Most countries have public lands that they designate for different uses. The following internationally recognized categories help countries classify land according to how people can (or can't) use it:

- ✔ **National parks:** Countries generally establish and manage national parks with the education and recreation of their citizens in mind. Often these parks are places of beauty or scientific interest (or both), and the resources within them (such as timber, minerals, and ore) can't be extracted. The purpose of some parks is to preserve certain species or ecosystems. National parks attract tourists, who can visit the park to be educated or to participate in recreational activities. The challenge with national parks is that tourists also increase the risk of environmental damage. Examples of national parks include Yellowstone in the U.S., Kruger in South Africa, and Blue Mountains in Australia.

- ✔ **National monuments:** National monuments are small areas of protected land, usually designated to protect a specific landmark (like the Arc de

Triomphe in Paris) or a unique cultural (Fort Sumter in South Carolina), natural (Sonoran Desert in Arizona), or geologic (Devils Tower in Wyoming) interest.

✔ **Managed resource protected areas:** Managed resource protected areas (MRPAs) are sections of land that have useful raw material resources (such as timber in forests) and that are carefully managed and protected to maintain those resources. In the U.S., national forests are an example of this type of land classification. Around the world, MRPAs provide raw materials and recreation while being cared for as public land.

✔ **Habitat/species management areas:** Countries protect and manage these land areas with the goal of preserving a certain species or a specific habitat. The management of these areas may include fire prevention, controlled burns, predator control, and regulated hunting. The overall goal of these methods is to maintain a certain biological community — one that often includes a species that's useful as a food or recreation resource, such as deer or fish.

✔ **Nature reserves and wilderness areas:** Nature reserves and wilderness areas are classified with restricted-use options in order to preserve and protect species and ecosystems. Unlike habitat/species management areas, nature reserves and wilderness areas are often left unmanaged and allowed to exist with as little human intervention as possible. These areas are often located within national parks or other protected and managed areas. This approach helps protect the innermost wilderness cores while allowing recreational use and resource management in surrounding areas. (I describe this concept further in Chapter 12.)

✔ **Protected landscapes and seascapes:** Many areas of the world are classified as protected and are used for tourism, recreation, and managed resource use. Often they contain unique or interesting species of plants and animals, but they can also include villages, orchards, beaches, and reefs.

Managing Land Resources

Environmental scientists present a few examples of land use management to illustrate the difference between unsustainable resource depletion and sustainable land and resource management. Two of the most common examples have to do with forests and grasslands, which I describe in the following sections. As a bonus, I also cover fire management, a type of resource management that relates to both forests and grasslands.

Timber! Harvesting the forest

A *forest* is any area of land that's more than 10 percent trees, including open forest savannahs, boreal forests, tropical forests, and temperate forests (see Chapter 7 for details on these ecosystems). The world's forests are one of the most valuable and most endangered resources.

Removal of trees, or the destruction of forests, is called *deforestation*. Throughout much of the developing world, humans clear trees from forests to use them as fuel — firewood — for daily life. In other regions, humans clear forests and replace them with agricultural or grazing land. This section explains two different methods for removing trees from forests — one unsustainable and one sustainable.

Clear-cutting forests

Clear-cutting, the most common method of timber removal, removes all the trees in a given area, often by using large machinery. From an economic standpoint, the benefit of this approach is that humans can harvest maximum timber with minimal energy and cost. However, clear-cutting destroys habitats and landscapes in the process. Figure 10-1 illustrates a patch of clear-cut forest visible as a scar on the forested landscape.

Removing all the trees from a large patch of land leaves the other organisms in the ecosystem without the shelter and resources the trees provided. The open patch increases the amount of sunlight that shines through to the forest floor and the temperature of the habitat surrounding the open patch, both of which affect the organisms that live there. Plus, large bare patches of clear-cut forest leave the soil vulnerable to erosion by wind and water.

Some timber harvesters replant clear-cut patches with a single species of tree. Although this replanting helps control the erosion problem, it doesn't reestablish the diverse ecosystem that was previously present in the forest habitat.

Selecting certain trees

Because people need to use the resources that forests provide (and because those resources aren't unlimited), environmental scientists have developed land management practices to help sustain forest resources into the future while using them to meet today's needs as well.

Selective cutting, also called *selective harvesting,* is a timber-harvesting approach that's less destructive than clear-cutting. This method of harvesting removes single trees or a few trees at a time from a forested area, leaving many behind. Selective cutting of trees requires more time, money, and energy to extract the individual trees, and it can still result in the soil erosion and ecosystem damage that are associated with other forms of timber harvesting (such as road building and air pollution). However, the disturbance to the ecosystem isn't as dramatic as it is with clear-cutting.

Figure 10-1:
A clear-cut
patch of
forest.

In an effort to find more sustainable ways of using timber resources, scientists and foresters in some places have developed methods of *ecologically sustainable forestry,* such as using pack animals instead of machinery to remove trees one at a time to reduce the need for roads and the impact on the forest ecosystem. The goal of this type of timber harvesting is to remove trees without damaging the surrounding habitat or organisms. Although this approach works well on a small scale, it isn't cost- or time-effective for the commercial-scale logging industry.

Grazing the grasslands

Grasslands are another ecosystem that people use (and sometimes abuse) for the resources they provide. Generally speaking, *grasslands,* or *rangelands,* as they're sometimes called, are open areas without trees. Farmers and ranchers use grasslands to graze animals such as cattle and sheep. Allowing herd animals to inhabit these open grasslands inevitably leads to some environmental damage.

Overgrazing is when herd animals consume the vegetation cover of a grassland to the degree that it can't be naturally replenished. After animals remove the vegetation by overgrazing, grasslands are susceptible to erosion by wind and water, which may convert the land into desert — thus making it generally unable to sustain vegetation. This process of changing previously fertile or useful land into a desert is called *desertification*.

One approach that minimizes ecosystem damage and desertification due to overgrazing is called *rotational grazing*. Rotational grazing requires that farmers and ranchers allow their herd animals to graze an area for only a short time and then move (or rotate) them into a different area. This rotation from one grazing spot to another mimics the natural process of herd animals grazing a landscape and allows the patches of grassland to recover before the animals return to graze them.

Questioning Smokey the Bear: Fire management

In both forested ecosystems and grasslands, land managers face difficult decisions about fire management. Wildfires are difficult to control and potentially destructive to human developments. With this potential destruction in mind, for most of the 20th century, land managers practiced methods of fire suppression, focused on keeping fires from starting and putting them out quickly when they did.

In recent years, however, ecosystem scientists called *fire ecologists* have recognized that fire is a welcome periodic disturbance in some ecosystems. For example, fires that sweep through some ecosystems help the processes of nutrient and matter cycling, and in some cases, species have evolved to depend on occasional fires. When land managers completely eliminate fire, those ecosystems and species suffer. (See Chapter 8 for a case study about ponderosa pine forests and fire adaptation.)

Both grasslands and forests may benefit from an approach to fire management that includes prescribed burns. *Prescribed burns* are fires that land managers set deliberately and control as they burn. The goal is to provide the ecosystem with the services of a fire without allowing the situation to grow out of control and threaten human lives or urban areas. Because many of the largest, most difficult to control fires are a result of the buildup of plant matter in the absence of fire, allowing occasional, controlled burns keeps the fires smaller and less intense and keeps the biomass, or burnable material, to a manageable level.

Considering the Connection between Urbanization and Land Use

As I mention in the introduction of this chapter, humans use land itself as a resource — as a space on which to build roads, cities, homes, and industry. Although many people still live in rural areas, where they depend closely on the natural resources in their ecosystem, the last 50 years have seen a surge in *urbanization* — or the development of cities around the world. The following sections take a look at how urbanization works and what people can do to help make urban growth more sustainable.

Sprawling across the horizon

In the U.S., most urban growth has occurred on land surrounding cities, called *suburban* areas. These regions support growing numbers of residential communities with access to the culture, goods, and services of a nearby city. As suburban areas grow to accommodate the larger populations that want to live in them, they create urban sprawl. *Urban sprawl* is suburban development that spreads into surrounding rural areas and transforms it into a landscape of homes, retail shops, and parking lots.

Some characteristics of a region that's experiencing urban sprawl include

- ✔ Low-density residential (single-family homes) and commercial development
- ✔ Dominance of freeways and automobiles on the landscape
- ✔ Numerous strip malls and parking lots
- ✔ No centralized planning or land use management
- ✔ Development that consumes farmlands and natural space
- ✔ Older housing in city centers that's left in disrepair or used to house low-income families

One of the reasons that people move to suburban areas and encourage urban sprawl is the lower cost of living in suburbs relative to living in a city. The seemingly endless supply of land outside the city keeps housing prices low as new development expands into previously rural areas. In contrast, living space inside the city is limited and, therefore, costs more.

As families and businesses spread into the suburbs, the downtown part of a city may experience urban blight. *Urban blight* is what occurs when residents move out of a city center into the more affordable suburbs; oftentimes, the city is left to deteriorate. Without a tax base to fund infrastructure and other community services, the city becomes a less desirable place to live and encourages even more relocation to surrounding areas.

The relationship between automobile-dependent transportation (from the suburbs to the cities and back again) and urban sprawl illustrates an interesting positive feedback loop, or chain of changes (see Chapter 6 for a description of feedback loops). The loop looks like this:

1. Governments use gasoline taxes to construct more better-quality highways.

2. With easy highway access to and from the city, residential suburban development increases.

3. Increasing suburban populations add to road congestion, leading to longer commute times into the city and more gasoline use.

4. Taxes collected from the increased use of gasoline go toward expanding and repairing highways, opening up even more land outside the city for the expansion of urban sprawl.

The cycle continues this way, feeding itself to the point that some people drive more than an hour from their home to the nearest city to work, and their entire commute takes them through sprawling suburban communities.

As urban sprawl continues, natural landscapes and rural areas around population centers disappear, taking with them the resources, recreation, and ecosystem services they provide. As a result, people must depend on resources from farther away (which require fuel and pollution to transport) and become less invested in the health of their own ecosystem.

Growing smarter

In response to concerns about how urban sprawl transforms natural landscapes into parking lots and adds to traffic congestion (and air pollution), some communities have begun to plan development in more sustainable ways. Well-planned development that uses existing infrastructure and land to maximize living space and conserve resources is called *smart growth*.

Smart growth isn't anti-growth or anti-development; rather, it's a development strategy that approaches the inevitable growth of urban areas with the intent of creating pleasant, healthy, and functional spaces for humans to inhabit. Smart growth includes limiting water and air pollutants and reducing, reusing, and recycling urban resources and spaces.

Some of the aspects of urban smart growth include

✔ Mixing residential, business, retail, and recreational land use

✔ Offering a range of housing options for families of all sizes and incomes

✔ Providing pedestrian-friendly neighborhoods, where people can walk to businesses and services (like the post office)

✔ Encouraging collaboration between everyone with an interest in the neighborhood when making decisions about development

✔ Building up instead of out (creating multistory buildings with ground-level business spaces and housing above)

✔ Fostering a sense of home so that members of the community take pride in living in and caring for their city or neighborhood

✔ Valuing and preserving natural spaces and local farmlands for the recreation, fresh produce, and ecosystem services they provide

✔ Building infrastructure around multiple transit options, including public transit (like buses and trains), shared car networks, and bicycling options

✔ Filling in existing urban space with redevelopment or repurposing abandoned lots in cities

✔ Making decisions about development that are fair and cost-effective

Chapter 11

Dishing It Up: Food Resources

*E*very living thing needs a fuel source to grow and function from day to day. Like other animals, people get their fuel by consuming food sources. The more people who live on Earth, the more the Earth's resources get stretched to provide food for them.

In this chapter, I describe the environmental issues related to food production, specifically farming. I explain what people need to do to maintain an environment healthy enough to produce food now and into the future, and I describe methods of food production that have degraded the environment, along with some methods that can help repair the damage. Although I focus on farming, I also briefly describe the challenges and environmental impacts of fishing and raising animals as a food resource.

Seeking Food Security

While some human communities have access to plentiful supplies of food, others struggle to eat enough to survive each day. *Food security* is having enough nutritional food to eat every day; unfortunately, not everyone in the world has it.

In this section, I explain the origins of famine and the nutritional needs that go along with food security. I also describe the industrialization of farming and the green revolution, which has led to increasing crop yields and the ability to feed more people, often at the expense of variety.

Identifying the factors that lead to famine

In extreme cases of *food insecurity,* or not having access to enough nutritional food every day, a society may experience a *famine.* Famines occur when people die due to lack of food. Scientists point to many different factors that lead to famine in parts of the world. Here are a few of the most common causes:

- ✔ **Food distribution is uneven.** The global environment provides food, or the resources to produce food, in quantities that are large enough for everyone to have food security. However, the distribution of those resources isn't spread evenly across the globe. For this reason, some environmental scientists think the biggest challenge to improving food security is figuring out how to better distribute available food resources.

- ✔ **Poverty limits access to food resources.** People in impoverished or poor communities around the world don't have the financial means or other resources needed to purchase or produce food for their families.

- ✔ **Political instability leaves refugee populations vulnerable.** In countries where the political system is unstable, families and whole populations sometimes find themselves displaced from their homes due to warfare. These people may have had access to food resources in their homeland, but after they relocate or are driven away from their homes and farms, they're left without access to a reliable food source.

- ✔ **Food supplies are fed to livestock rather than people.** Nearly 40 percent of the grain grown in agricultural fields is used to feed cattle and chickens. Although these animals then become sources of food for some people, a large amount of energy gets lost in the process as a portion of grain that could feed many people is transformed into beef or poultry that feeds only a few. I cover this issue in more detail later in the section "Farming Fish and Other Animals."

- ✔ **Human populations have exceeded the ability of agriculture to support them.** Although the production of crops has increased (in particular grains such as corn, rice, and wheat), so has the number of people being supported by these crops. Some environmental scientists even suggest that humans are close to exceeding their carrying capacity (see Chapter 8 for an explanation of carrying capacity).

Eating for health

Providing food security isn't as simple as producing enough crops to meet a daily caloric minimum. To be healthy, humans need to eat a variety of food types to get different nutrients in their diet. A wide range of plant and animal resources can help people meet their daily nutritional requirements.

People who lack access to basic nutrients suffer from *malnutrition* — an imbalance in the vitamins, minerals, proteins, and other nutrients needed by the human body. People who are malnourished may consume enough calories each day, but the calories they eat don't contain all the different elements needed for a balanced and healthy diet. Hence, they suffer from *nutrient deficiency*. Different nutrient deficiencies lead to different severe health conditions. Table 11-1 lists some of the most common nutrient deficiencies and the health problems they can cause.

Table 11-1	Common Nutritional Deficiencies and Corresponding Health Conditions
Nutrient Deficiency	*Health Condition*
Iron	Anemia (low red blood cell count)
Iodine	Goiter (swollen thyroid)
Vitamin A	Blindness
Folic acid	Neurological problems
Protein	Kwashiorkor
Protein/calories	Marasmus

In the last 20 years, *overnutrition,* or the eating of too many unhealthy foods, has become more common than it used to be. Overnutrition results in dangerous health issues, including heart disease, high blood pressure, and diabetes (a disease in which your body doesn't properly process sugars).

The key to healthy and nutritional eating is having access to a wide variety of food resources. In the past, humans used more than 3,000 different species of plants as food; today, only 16 species represent most of the food produced by large, industrialized farms, greatly reducing the biodiversity of food-producing ecosystems (see Chapter 12 for details on why biodiversity is important).

Feeding the world: The green revolution

Farming, or growing plants as a food resource, isn't as simple as it used to be. When humans first started planting, plowing, and harvesting crops nearly 5,000 years ago, all a farmer needed were a few tools and a willingness to get his (or her) hands dirty. More recently, particularly over the last 100 years or so, farming has slowly transformed into industrial agriculture or *agribusiness.*

At the beginning of the 20th century, most farmers operated their farms to feed and financially support their family or community. By the end of the 20th century, however, farms had grown into a massive industry capable of feeding the world. The result of the increased use of industrial technology in farming led to a revolution of sorts — *the green revolution,* to be exact — producing huge amounts of food from the land. In this section, I describe the components of the green revolution in agriculture, including the pros and cons of this relatively new way of growing crops.

Increasing crop yields

The green revolution has incorporated technological advances in machinery and chemical production to increase the number of crops harvested each year, which is known as the *crop yield.* This increase in crop yield is the result of four important changes in how farmers produce food crops:

- **Mechanization:** Most modern farmers do nearly all aspects of industrial crop production on a massive scale, using large machines. From plowing, planting, and harvesting to sorting, drying, and preparing for market, the green revolution has mechanized the farming process.

- **Irrigation:** Agribusiness farmers construct large and complex systems to bring water to the crops instead of depending on seasonal rains or rivers. New methods of irrigation have expanded farms into areas that previously couldn't have supported agriculture.

- **New crop varieties:** Through breeding and genetic modification (more on this later in this chapter), scientists have developed new varieties of food crops. These new varieties are called *high-responder crops* because they grow quickly and produce a high yield in response to the proper nutrients and water conditions.

- **Fertilizers and pesticides:** The development of *synthetic,* or manmade, fertilizers and pesticides, coupled with the development of high-responder crop varieties, results in maximum amounts of food being produced each season.

Weighing the costs of industrialization

The result of the technological advances in agriculture that I describe in the preceding section is the ability to meet the growing need for food of a growing human population. But while meeting the needs of the world's growing population is clearly a benefit of agricultural industrialization, it also has its share of drawbacks. I describe a few of the main ones here:

- **Monoculture ecosystems:** Large farms that work with industrialized systems of plowing, planting, and harvesting maximize the efficiency of these machines by planting expansive fields of a single type of crop, thus creating a *monoculture ecosystem.* In general, monoculture ecosystems, by nature, aren't as resilient to pests and other disturbances as more biodiverse and complex ecosystems (see Chapter 12).

- **Increased energy use:** The energy invested to produce and transport crops has increased dramatically over the last 50 years. For one, you need fuel sources to run the machinery of an industrialized farm. Plus, manufacturers use fossil fuels to create nitrogen-rich fertilizers necessary to support one or two types of crops and high-responder varieties.

- **Higher expenses:** Machinery, fuel, pesticides, and fertilizers cost money. So although high-responder varieties of certain crops promise high yields, purchasing seeds for them also costs more. Because the seeds are modified or specially bred, farmers have to purchase them from the seed companies who develop them.

- **Pollution:** Burning fossil fuels to power farm machinery pollutes the air. Adding fertilizers and pesticides to fields means that they'll eventually wash into nearby streams, rivers, and other water supplies. (I provide details on the dangers of pesticides in Chapter 17 and the effect of fertilizers, or *nutrient pollution,* in Chapter 16.)

- **Increased erosion:** Using large machinery to clear, plant, and plow large plots of land creates conditions that are ideal for erosion, which I describe in the next section.

Digging in the Dirt: Components of Healthy Soil

To grow crops for food, you need land that's fertile, or *arable.* For land to be arable, it needs water, sunlight, and healthy soil. In this section, I explain what it takes to maintain healthy soil that will support crops, and I describe how advances in farming technology reduce soil fertility and what farmers can do to fix the problem.

Learning the ABC's of soil

If you think soil is simply dirt, think again! *Soil* is a complex blend of many different parts or components. In fact, a handful of soil is an ecosystem of its own — a balance of living and nonliving elements working together — where energy is transformed and matter is recycled. Figure 11-1 illustrates one example of what a healthy soil ecosystem looks like.

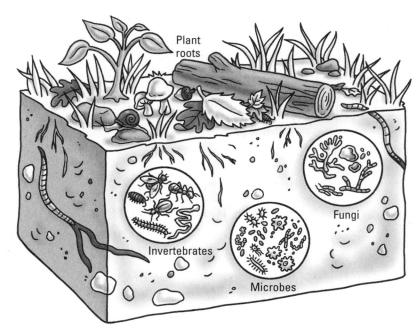

Plant roots

Fungi

Invertebrates

Microbes

Figure 11-1:
Components
of a healthy
soil
ecosystem.

Illustration by Lisa Reed

Healthy soil includes sediments of different sizes (such as clay and sand), decaying organic matter, living organisms (such as bacteria, worms, beetles, and other decomposers), water, and air. The absence or overabundance of any one of these soil components can cause big problems for a farmer or gardener trying to grow plants.

Scientists examine soil in columns, called *soil profiles*. Each layer of a soil profile has specific characteristics that define it. Figure 11-2 illustrates a soil profile with the major layers or horizons that soil scientists recognize; I explain each of these layers in the following list:

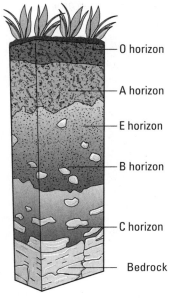

Figure 11-2:
Soil
horizons.

Illustration by Wiley, Composition Services Graphics

O horizon

A horizon

E horizon

B horizon

C horizon

Bedrock

✔ **O horizon:** The top layer of soil consists of surface litter, or decaying plant material, insects, and other invertebrates. This layer is called the *O horizon*.

✔ **A horizon:** The *A horizon* is the layer where plants are rooted and is sometimes called the *topsoil*. It consists of sediments mixed with organic matter, water, and air — all the components needed to support plant growth.

✔ **E horizon:** Below the A horizon is the *E horizon*, which contains less organic matter and less air than the A horizon. In this layer, water dissolves minerals and organic compounds and carries them along as it moves downward into deeper layers of the soil. Because these minerals and compounds move through this layer, the E horizon is sometimes called the *zone of leaching*.

✔ **B horizon:** The minerals and compounds that move down through the E horizon accumulate in the next layer, the *B horizon*. This layer, also called the *subsoil*, is rich in the minerals and tiny particles that flow downward from the upper soil horizons.

✔ **C horizon:** Near the bottom of a soil profile, far from the topsoil, is the *C horizon*, which is composed mainly of mineral components and pieces of the bedrock below it. This layer is the least fertile because it lacks the air and organic matter that are more common in the upper layers.

✔ **Bedrock:** Below the C horizon is the *bedrock*. Bedrock may be any type
of rock (see Chapter 13 for details) and varies from region to region. The
type of rock and the minerals in the rock play a role in determining what
type of plants may grow in the soil above; however, the characteristics
of the bedrock aren't nearly as important as the organic matter, air, and
water found in the upper soil horizons when you're trying to grow plants.

Considering erosion: The removal of topsoil

The uppermost layer of soil, the topsoil, is the most important in terms of
soil fertility, but it's also the most susceptible to *erosion,* or removal by wind
and water. Plants help anchor the soil in place and keep it from washing or
blowing away. Every time farmers harvest their crops, they leave the soil vulnerable to erosion. Soil erosion occurs in many different ways. Here are some
of the most common ways water erodes croplands:

✔ **Sheet erosion:** *Sheet erosion* occurs when water washes across a level
soil surface, washing away a thin layer, or *sheet,* from the topsoil.

✔ **Rill erosion:** *Rill erosion* occurs when tiny streams or rivulets of water
create little channels, or *rills,* in the soil surface as they wash the
topsoil away.

✔ **Gully erosion:** *Gully erosion* looks similar to rill erosion but on a much
larger scale. As the water washes soil away, the water cuts large channels through the soil surface.

Through the processes of sheet and rill erosion, water removes small
amounts of soil a little bit at a time, but over a season, these small amounts
add up to quite a bit of topsoil lost.

Wind erosion also removes topsoil. After farmers or grazing animals remove
the vegetation from fields, wind can easily blow away the topsoil from the
surface. After the topsoil is gone, nothing can support new plant growth. The
result is a condition called desertification.

Desertification is when previously fertile land transforms into desert, unable to
support plant growth due to overuse or misuse. Unfortunately, the intensive,
industrial farming practices of the last few decades have led to increasing
desertification of arable lands.

The Pitfalls of Pesticides

One of the major contributions of technology and industry to farming has been the development of chemicals to help minimize insect damage and weed competition in agricultural fields. In this section, I describe the various types of pesticides used to increase food production, as well as the dangers of depending on such chemicals (particularly synthetic or manmade chemicals) to produce the food you eat.

Concocting a formula for every pest

Farmers encounter many different pests that can damage their crops, and each kind of pest requires different chemical compounds to kill it. Table 11-2 lists a few examples of different types of pesticides and what pests they target.

Table 11-2	Different Types of Pesticides
Name	*Target*
Herbicide	Plants (weeds)
Insecticide	Insects
Fungicide	Fungi (mushrooms)
Biocide	Any living thing

The use of pesticides isn't new, but these days most of the pesticides farmers use are synthetic rather than *organic,* or natural. Scientists create synthetic pesticides in a laboratory and can engineer them specifically to target a particular problem pest. While this control seems like a good thing, it results in a never-ending battle between farmers and pests and the need to constantly develop new chemical formulas, as I explain in the next section.

Running in circles: The pesticide treadmill

Agricultural pests, like all other organisms, strive to survive. When farmers expose them to a chemical compound that's designed to kill them, a few of them invariably survive to reproduce. As a result, the next generation is likely to be resistant to the deadly chemical (after all, their parents had some genetic mutation that allowed them to survive; check out Chapter 12 on biodiversity and evolution for details).

Thus begins a cycle some scientists call the *pesticide treadmill:* As farmers and pesticide producers work harder to control pests, they create pests that are harder to control. The pesticide treadmill starts with the development of a new and powerful pesticide, but once it starts, it never ends. Here's how the cycle goes:

1. Chemists create Pesticide A, which is extremely effective at killing pests.
2. Farmers apply the pesticide to their crops.
3. Most pests die, but a few survive and reproduce.
4. A new generation of pests that aren't affected by Pesticide A is born.
5. Farmers apply Pesticide A with little or no result now that the pests are resistant.
6. Chemists create a newer, more powerful compound, Pesticide B.
7. The cycle repeats itself over and over again.

The pesticide treadmill leads to increasing costs for farmers, who have to purchase new pesticide formulas every year or risk losing their crops. Plus, each generation of pests is increasingly more difficult to kill, meaning that the pesticides chemists develop to kill the new pests must be even more potent than the last ones. And, as I describe next, spreading highly effective poisons around an ecosystem can have disastrous results.

Spreading poisons far and wide

Not all pesticides are manmade. Some, such as botanical extracts and oils, are naturally occurring compounds. Naturally occurring pesticides are still extremely toxic and can affect both the humans and the pests that are exposed to them. However, naturally occurring pesticide compounds eventually break down and recycle back through the ecosystem, whereas manmade compounds often do not.

Scientists categorize synthetic or manmade pesticides according to their chemical structure. Here are a few of the most common types and the effects they have on their environment and the living things in it:

- ✔ **Organophosphates:** *Organophosphates* are the most common and widely used synthetic pesticides. Organophosphate compounds break down eventually, but they're highly toxic for the first few days after they're applied. These chemicals are related to the nerve gases used in World War II, and they can result in severe damage to the nervous system (brain) of animals, including humans.

- ✔ **Chlorinated hydrocarbons:** *Chlorinated hydrocarbons* are also called *organochlorines.* These compounds don't easily break down and, therefore, remain in the environment long after farmers apply them to their cropland. Chlorinated hydrocarbon compounds, such as the commonly used Atrazine, have been linked to cancer, are considered an endocrine disruptor (see Chapter 17), and are banned in Europe.

- ✔ **Inorganic pesticides:** *Inorganic pesticides* are chemical compounds that don't include carbon molecules. (In case you're wondering, all the other pesticides in this list do include carbon molecules.) Inorganic pesticides usually contain toxic elements, such as mercury and arsenic. These elements remain in the soil and ecosystem long after they've been applied to crops and enter the food chain. Check out Chapter 17 for details on the effects of toxins such as mercury and arsenic.

Dangerous organic compounds that remain in the environment without breaking down are called *persistent organic pollutants* or *POPs.* (In this case, the term *organic* simply means that these compounds are built of carbon molecules, which is different from the use of the word *organic* to describe food; see the later section "Defining Organic" for details.)

POPs, such as those found in the synthetic pesticides in the preceding list, exist for years, even decades in lethal amounts. They remain in the soil or are washed into water supplies or groundwater, where they circulate into every corner of the environment. Some of them even evaporate into the atmosphere and then rain down into a completely different environment, far from the farmlands they were originally applied to.

POPs are particularly dangerous to animals and humans because they can accumulate in living organisms (through a process called *bioaccumulation*) or become increasingly concentrated up the food chain (a process called *biomagnification*). For more details on POPs, bioaccumulation, and biomagnification, turn to Chapter 17.

"Frankenfood": Genetically Modifying Food

One modern approach to overcoming the challenges of producing enough food to meet the needs of the Earth's growing population is to change, or *modify,* plants at the genetic level. It may sound like science fiction at first, but *genetic modification,* or *genetic engineering,* of plants is more common than you think. In fact, you've probably eaten some genetically modified food today; more than 60 percent of processed food in the U.S. contains genetically altered corn or soy. In this section, I explain how and why scientists genetically modify plants and animals, and I dive into a heated debate about *genetically modified organisms,* or *GMOs.*

Splicing and dicing chromosomes

To change the genes of an organism, scientists cut part of the DNA out of one organism and insert it into the DNA of another organism through a complicated process. In doing so, the scientist can select a specific genetic trait from the first organism and add it to the other organism, even if the organisms could never or would never combine their DNA in the natural world. Organisms that contain genes from another species are called *transgenic.*

Modifying plants

Genetic modification works in both plants and animals, but currently it's most common in plants with the goal of increasing crop yield. The three main ways that scientists engineer plants toward this goal are

- **Making plants resistant to pesticides and able to repel other pests:** One way to increase crop yields is by creating plants that are resistant to pesticides. In fact, these plants were the first widespread genetically engineered crops that scientists created. Thanks to this resistance, these genetically modified crops can survive pesticide applications, while all the weeds around them die. Similarly, scientists also engineer crops that repel insects by producing natural pesticides. When farmers use these crops, they don't have to apply synthetic pesticides to the cropland itself.

- **Increasing plants' hardiness:** Another way to increase crop yields is by modifying plants to survive and thrive in conditions that are less than perfect. For example, scientists create crops that grow well in soils with low nutrients or water with too much salt.

✔ **Increasing the growth rate of plants:** A third way to increase annual crop yields is to create crops that grow from seed to maturity more quickly. With such crops, farmers can complete multiple harvesting cycles in one year, greatly increasing the output of their agricultural land.

Other ways that scientists genetically modify plants include adding nutrients or vaccines and removing allergens. Although the use of genetically modified plants to increase food supply and add nutrition looks like progress, the safety of genetically engineered food is still up for debate among scientists.

Modifying animals

Much less common than genetically modified plants but still a growing area of research is the genetic modification of animals for food. Two examples of animals that have been genetically modified are cattle and fish:

✔ **Cattle:** Scientists have developed cattle species that are resistant to some diseases or that produce milk or meat with higher levels of protein.

✔ **Fish:** Scientists have modified different farmed fish species, including salmon, catfish, and tilapia, to mature more quickly, thus increasing production numbers.

Although the potential for positive outcomes with the use of genetically modified animals is out there, many questions about the effects they will have on ecosystems and human health remain unanswered. Whether the benefits (namely, being able to feed a growing human population) outweigh the dangers remains to be seen.

Debating GMOs

Today, scientists are in the middle of a strong debate about the safety of consuming genetically modified organisms (GMOs). Like any debate, this one has two sides.

The pro-GMO side lists the following benefits of using GMOs as a food resource:

✔ Reduced use of chemical pesticides that damage the environment

✔ Increased crop yields and animal production, which lead to greater food production and increased food security around the world

✔ Increased efficiency in the use of soil, water, and land resources to produce food

✔ Increased quality of food, including more nutrients, a longer shelf life, and fewer allergens

The anti-GMO side lists the following concerns about using GMOs for food:

✔ How GMOs can affect the health of the animals (including humans) that consume them is still unknown.

✔ In the case of both plants and animals, some people are concerned that modified genes will enter native (nonmodified) populations and impact entire ecosystems in ways that can't yet be predicted. For example, the possibility exists (and has been observed in a few instances already) that engineered genes will transfer through natural pollination to weeds, creating *superweeds* that are even more difficult to control.

✔ Ingestion of GMOs with pesticide and disease resistance may not be healthy for other organisms in the food web; as a result, GMOs can cause disruptions or ecosystem collapse.

✔ No one knows how the transfer of distantly related or unrelated genetic material from one organism to another will play out in the long-term cycle of natural selection and evolution after GMOs are released into the environment.

One solution for dealing with the safety concerns of consuming GMOs is to have food producers clearly label any food product that contains transgenic organisms and then let consumers make the choice about what they consume. In the U.S., no government agency (at the federal, state, or local level) has declared that manufacturers must label products containing GMOs. Regulating agencies have taken an "innocent until proven guilty" approach concerning GMOs instead of exercising the precautionary principle (I explain each of these approaches in Chapter 20). Currently, companies can voluntarily label their products to describe what, if any, GMOs are included, but they don't have to. Only time will tell how GMOs affect ecosystems and human health.

Considering the ethical issues related to GMOs

Aside from the effects GMOs may have on the health of humans, other animals, and the environment, scientists and farmers also have to consider the business ethics issues related to GMOs. Central to these issues is the fact that the corporations that develop GMOs are the legal owners of the products they create and the knowledge of how to make them (this knowledge is considered *intellectual property*).

GMO crop seed has already been the focus of legal and ethical dilemmas (dilemmas concerning GMO animals are certainly on the horizon). The corporations that develop genetically modified seeds are the only source, worldwide, for farmers who want to purchase and plant specific transgenic crops. The two main problems with this setup are

✔ **Inequality among farmers:** The expense of purchasing transgenic seeds from large corporations means that some farmers, particularly those who work on small farms in developing nations, don't have access to these seeds. As a result, they can't compete in the free market against larger farms that do have access to pesticide-resistant, quick-maturing, high-yield crops.

✔ **Liability for natural pollination:** As soon as farmers plant the transgenic crops, they freely produce pollen to reproduce and may cross-pollinate with similar species on neighboring cropland. The seed produced by this cross-pollination may retain some of the transgenic DNA. Seed companies have made the case that any plant containing genetic material engineered in their laboratories remains the property of the corporations. In some instances, this issue has led to lawsuits, in which corporations sue neighboring farmers for piracy or stealing of their intellectual property, when in reality, it was just blowing in the wind!

Farming Fish and Other Animals

Not all people are *vegetarians,* or people who eat only plants. In fact, most people eat some form of animal protein as well. Similar to the way modern technology has turned small farms into larger agribusiness operations, that same technology has revolutionized *animal husbandry,* or the breeding and raising of animals for food.

In this section, I describe how depending on animals for protein is less energy efficient than eating plants, and I cover the advances in meat production that have increased the availability of protein as a food source. I also explain the effects that these advances in mass meat production have on the environment and the living things in it.

Eating less efficiently

Animal meat and other animal-related products, such as milk, cheese, and eggs, are important sources of protein for many people. Compared to plants, however, meat is expensive and inefficient to produce. For one, the modern technology used to produce animals in larger numbers for human consumption requires a lot of fossil fuel energy. Consider that you need more than 15 times the amount of fossil fuel energy to produce 1 kilogram of beef than you need to produce a similar amount of vegetables.

Another important point to consider is that animals are consumers (not producers), and consumers have to eat enough plants to function in daily life, grow, and reproduce. As I explain in Chapter 4, each time energy is transferred up the food chain, some of that energy is lost.

Try thinking about animal production this way: To produce 1 kilogram of consumable beef, a cow must eat 8 kilograms of grains. Thus, you could simply eat 1 kilogram of the original grain yourself (as bread) and save everyone money and time in the process.

Confining animals for mass production

One way to quickly mature, fatten, and reproduce animals such as cows, pigs, and chickens is in a large operation where breeders keep animals together in large enclosures and feed them special fattening feed. These confined animal feeding operations, or *CAFOs* as they're often called, are used in cattle and chicken husbandry regularly. Like large-scale crop production, this large-scale animal production creates several environmental and health issues, including the following:

- **High disease rates:** Animals in CAFOs eat a mix of corn, soy, and protein that encourages maximum growth in minimum time. The faster they grow and mature, the more quickly their owners can sell them as food. Unfortunately, thousands of animals are often confined in a single barn or warehouse, and keeping so many animals in those conditions creates numerous health hazards. One such hazard is that diseases can spread quickly among the confined animals.

  To reduce the spread of disease, much of the feed also includes regular doses of antibiotics as a preventive measure. So instead of treating animals when they get sick, operators use the medicated feed to ward off potential illness. This liberal use of antibiotics leads to the health issue in the next bullet.

- **Development of antibiotic-resistant, disease-bearing microorganisms:** The use of antibiotics in the animals' feed leads to antibiotics in the meat that's produced as well as in the manure that accumulates and washes into nearby water supplies. Spreading antibiotics throughout the environment creates conditions for disease-bearing microorganisms to evolve a resistance, creating *superbugs* that require stronger antibiotics in both animals and humans. This cycle is similar to the pesticide treadmill that I describe earlier in this chapter.

- **Increased water pollution from animal waste:** Large operations with hundreds of animals mean literally tons of fecal waste that must be properly disposed. When it rains and animal waste washes into nearby waters, you end up with dead zones, eutrophication, and the spread of pathogens (see Chapter 16 for details on water pollution).

Farming fish

Cows, pigs, and chickens aren't the only animals humans depend on for protein in mass quantities. Humans also harvest fish of all kinds, using technologically advanced and efficient methods to maximize food production.

For years, people have used large nets to capture wild fish for food. Unfortunately, this mass fishing technique has led to dramatic declines in fish populations and a disruption of ocean food chains. (Turn to Chapter 23 for a description of overfishing in the Grand Banks of Newfoundland.) One solution people have come up with is to farm fish in pens, where they can grow, fatten, and reproduce quickly. These fish pens are similar to the CAFOs for cows, pigs, and chickens that I discuss in the preceding section. The growing of fish in pens, netted lagoons, or tanks — called *aquaculture* — results in many of the same problems that crowded CAFOs do, including disease, overproduction of waste, and environmental pollution.

Fish farmers have come up with some interesting approaches to overcome the problems related to aquaculture. One such approach is creating aquaculture tanks that include a biodiverse ecosystem, consisting of organisms at different trophic levels that fill different ecological niches (see Chapter 8). These *integrated polyculture systems,* as they're called, are most common in China, but they illustrate that more ecologically sound methods of fish farming are possible.

Establishing Sustainable Agriculture

Not all farming methods have harmful effects on the environment. In fact, farmers and scientists have developed several methods to produce food that don't damage the environment or rely on toxic chemicals. These methods are considered *sustainable agriculture* because they produce food in ways that keep ecosystems and farmland healthy enough to continue to produce food into the future. In this section, I describe a few different methods of farming, including pest control, that minimize environmental damage.

Conserving soil

Fertile soil is literally the foundation of farming. As I describe earlier in this chapter, fertile topsoil is vulnerable to erosion, particularly after the crops that grow in them have been harvested. The following methods of working the land help keep soil healthy and reduce the effects of wind and water erosion:

✔ **Agroforestry:** *Agroforestry* is the planting of trees and crops together across a field. The trees protect the topsoil by anchoring it in place and blocking winds that could blow it away. And, of course, farmers who use agroforestry can also harvest resources such as firewood or fruit from the trees.

✔ **Contour plowing:** Soil moves downslope easily with a little help from gravity and water. One way to keep soil in place is to plow in contours that follow the shape of the land. *Contour plowing* creates crop rows across sloped hillsides rather than up and down the slopes. These rows keep soil from washing downhill.

✔ **Intercropping:** *Intercropping* or *strip farming* is a method of planting two different types of crops in alternate rows in one field. If you harvest the crops at different times of year, then some vegetation is always present to anchor the soil.

✔ **Crop rotation:** Farmers who use *crop rotation* plant different crops in their fields each year for a few years in a rotating cycle. This keeps the soil healthy, especially if some years the crops are nitrogen-fixing plants that help replenish nitrogen in the soil. Crop rotation also breaks the reproductive cycle of pest species who often prefer a particular crop (I provide details on how this works in the next section).

✔ **Terracing:** Farms that use *terracing* create step-like or shelf-like patterns across sloping hills. Each terrace provides a narrow, flat space to plant crops and retains the soil that would otherwise wash downhill.

✔ **Reduced tillage:** In most cases, the worst soil erosion occurs when fields are *tilled,* or plowed, to prepare the topsoil for new seeds. *Reduced tillage* involves plowing tools that prepare the soil without removing all the vegetation or without churning up too much of the soil. Reduced tillage often requires the use of herbicides to help control weeds since some vegetation is left in the fields.

No matter which sustainable agriculture methods farmers use, they also have to add nutrients and organic matter to the soil regularly. In some of the methods in the preceding list, the vegetation left on the fields is a source of organic matter and nutrients. Farmers can also spread natural fertilizers such as manure, across their fields. For farmers who work with both animals and crops, this is a great way to recycle what would otherwise be a waste product (manure) into something beneficial.

Integrating pest management

In an attempt to reduce the use of pesticides, some farmers practice *integrated pest management,* or *IPM.* Instead of using a single method of pest control, IPM combines practices of crop rotation and intercropping with minimal chemical use. The goal of IPM is to reduce the use of chemical pesticides by fostering a habitat that discourages pests.

If an insect or disease infects a *monoculture crop* (a crop composed of a single species), the entire harvest is in danger of infection. Using crop rotation and intercropping helps reduce this type of damage. Many pests prefer to eat a certain species, and rotating the crop or intercropping interrupts the continuous food supply that a monoculture crop provides. Another advantage to intercropping is that farmers can plant something that attracts *prey insects* — insects that will prey on the pests and keep their populations under control naturally.

IPM techniques do rely on chemical pesticides when necessary, but usually only as a last resort. By trying other methods first, the need for chemicals greatly decreases. This decrease in the number of chemicals being used on crops not only keeps pollutants out of the environment but also saves the farmer the cost of continually purchasing stronger pesticide formulas. The highest cost in IPM comes upfront in training farmers how to manage, care for, and maintain their crop-growing ecosystems.

Integrating garden pest management

Think integrated pest management doesn't apply to you if you aren't a farmer? Think again! You can apply the principles of integrated pest management to gardens as well as farms. Regardless of how big your garden is, you probably encounter weed and insect problems. Here are some of the U.S. EPA's recommendations for reducing the use of pesticides in your garden:

✔ **Identify the culprits.** Determine what kinds of pests you have in your garden. Local garden stores or the county cooperative extension may be able to help you do so.

✔ **Set a goal.** Decide what level of pest management you're willing to undertake. Perhaps replacing pest-sensitive plants with hardier species is one solution. Or perhaps having a few weeds in the yard is okay with you.

✔ **Cut off supply lines.** If possible, remove the food, water, or other pest attractions from around your home and garden.

✔ **Think outside the grass.** In some regions, grass is difficult to maintain without herbicides for weed control. Think about other types of groundcover that you could use in place of grass.

✔ **Spend time on soil.** Keep your soil healthy so that the plants you want to grow have a fighting chance against weeds and insects.

✔ **Encourage pest predators.** Ladybugs, spiders, wasps, and ants are common insects that feed on other insects and not on your plants.

✔ **Get your hands dirty.** Pulling weeds and cultivating your garden by hand keeps you aware of what pests are present and which plants are suffering.

✔ **Apply chemicals with care.** If you decide to use a chemical pesticide, use the least amount that may be effective and be sure to read all instructions about safety, storage, and disposal.

✔ **Observe and assess.** Consider the results of your various methods and determine whether the risk of using chemicals is worth their effect on the pests.

Rice farmers in Indonesia provide an excellent example of the benefits of IPM. In the mid-1980s, Indonesian rice farmers used more than 50 different chemical pesticides. Studies showed that farmers sprayed these chemicals on their crops a few times a week, more out of habit than out of need to remove pests. Then the president of Indonesia banned most of the pesticides and started a program to educate rice farmers about IPM. Within two years, Indonesian rice farmers who practiced methods of IPM, such as inspecting their crops, allowing natural predators to help control pests, and spraying chemical pesticides only as a last resort, were bringing in higher crop yields than farmers still practicing indiscriminate pesticide application. With such success, IPM practices quickly spread throughout Indonesia, replacing previous methods of pest control.

Defining Organic

The term *organic* has taken on many different meanings in the last few decades. Fortunately, now that it's being applied to food labels, regulatory agencies have taken the time to define it.

According to the U.S. Department of Agriculture (USDA), foods that are *certified organic* must be grown and processed without synthetic pesticides or fertilizers. Animals must be fed organic feed, raised on free-range farms (not confined in cages), grown without steroid growth hormones or genetic modification, and given antibiotics only as treatment for illness (not as a preventive measure).

Of course, few things with the government are ever black and white, and the organic label is no different. In terms of food labels, there are many shades of gray, or different degrees of organic:

- **100% organic:** To legally have this label, the raw, fresh, or processed foods must contain only organic products and processing agents (except for water and salt, which are, by definition, inorganic compounds).

- **Organic:** A food labeled simply *organic* must have at least 95 percent certified organic ingredients (except for water and salt).

- **Made with organic ingredients:** Food products with this label must be made with 70 percent organic ingredients. The labels on these products often list exactly which of the ingredients are organic.

Products that have fewer than 70 percent organic ingredients are encouraged to list each organic ingredient separately on their labels, but they can't use the USDA organic seal on their packaging.

Reclaiming and replanting city blocks

The middle of a city may be the last place you'd expect to find rows of corn, beans, kale, and tomatoes. However, residents of the largest urban centers in the U.S., such as Los Angeles, Detroit, and New York, have begun to create farmland on abandoned and vacant city blocks. (You know the type — land surrounded by chain-link fences, covered in scraggly weeds, and littered with windblown garbage.) Transformed into vegetable gardens, these previously unused spaces in the city are now being put to good use, growing fresh food for local citizens. Community members tend the gardens and then distribute the crops throughout the community or sell them at local inner-city farmer's markets.

Many scientists agree that urban farms are exactly what sustainable cities need because they help accomplish the following goals of sustainability:

✔ To make use of unused and abandoned urban land

✔ To increase community food security

✔ To provide access to fresh, healthy, local, and in many cases, organic produce

✔ To bring community members together to work for their own benefit

✔ To transform concrete and pavement into green space

✔ To improve the overall environmental health of the community

Look around your own neighborhood and you may find a few empty lots or some unused land that could be used to grow food or transformed into green space that would benefit everyone in your community!

Chapter 12

Greater Than the Sum of Its Parts: Biodiversity

. . . .

In This Chapter

▶ Defining biodiversity

▶ Understanding evolution by natural selection

▶ Identifying the main threats to biodiversity

▶ Considering different ways to conserve biodiversity

. . . .

*T*ake a look around and you'll see that you inhabit a world filled with variety. The Earth is a patchwork of different ecosystems, each filled with various species. A single species alone may seem like a small thing, but when you combine many species together, they become greater than the sum of their parts: They become a healthy, functioning ecosystem that sustains life on the planet (see Chapter 6 for details on ecosystems).

Human survival depends on the availability of natural resources, which thrive in healthy ecosystems. In general, ecosystem health is related to ecosystem diversity. In this chapter, I explain how scientists define and study biological diversity, and I describe how this diversity comes about — evolution by natural selection. I also cover the many ways in which humans benefit from maintaining the diversity of organisms and ecosystems around the world, and I describe different approaches to conserving and protecting that biodiversity.

Variety 1s the Spice of Life: Defining Biodiversity

Biological diversity, or *biodiversity,* is the variation of living things on Earth. Environmental scientists observe this variation at the following different levels or scales:

✔ **Ecosystem diversity:** *Ecosystem diversity* is the variation in ecosystems in a particular region or location. It's the largest scale of biodiversity. A region with high ecosystem diversity has many different ecosystems, such as forests, grasslands, wetlands, and lakes, within it.

✔ **Species diversity:** *Species diversity* is the most commonly recognized scale of biodiversity. Species diversity is the variety and abundance of different species living in an ecosystem. An ecosystem with high species diversity has many different species coexisting within it. For example, lowland rainforests in Africa and South America may contain thousands of species in each square mile.

✔ **Genetic diversity:** *Genetic diversity* is the most specific level of biodiversity; it looks within a species at the variety of genes found in a population. The higher genetic diversity a species has, the wider range of physical and other characteristics that species has to adapt to environmental change.

Preserving biodiversity at all levels (ecosystem, species, and genetic) is important because it enhances the resilience of life on Earth. High ecosystem diversity gives people and other organisms a wide array of resources to use. A large variety of species creates complex food webs and other relationships that strengthen an ecosystem's resilience to disturbance. And a high amount of genetic diversity means that new and unique traits are appearing that may be adaptive and useful under certain conditions. As global climates change and human urbanization continues, the ability to adapt will become even more important to plant and animal species, including humans.

Examining species richness and evenness

Scientists most often study and calculate biodiversity at the level of species diversity. To measure species diversity, scientists could simply count the number of different species on Earth. However, organisms aren't spread across the Earth evenly; some places have higher species diversity than others. To get a more accurate picture of the patterns of species diversity, scientists measure species richness and species evenness:

✔ **Species richness:** *Species richness* measures the number of different species in a given area or ecosystem. If an ecosystem has a lot of different species, then it has high species richness.

✔ **Species evenness:** *Species evenness* measures the relative abundance of each species. If an ecosystem has a similar number of individuals in each species, then the ecosystem has high species evenness.

Here are a few examples of different levels of species richness and evenness:

✔ If you have 4 people, 1 dog, 3 goldfish, and 2 cats in your house, you have high species richness but low species evenness.

✔ If you have 2 people, 2 dogs, 2 fish, and 2 cats in your house, you have both high species richness and high species evenness.

✔ If you have 2 people and 2 dogs in your house, then you have low species richness and high species evenness.

Benefitting from biodiversity

Biodiversity isn't just a scientific concept or measurement. Variety in living organisms and ecosystems is important to humans as well. Here are some of the most important ways that humans benefit from biodiversity:

✔ **Biodiversity provides food and medicine.** High levels of ecosystem, species, and genetic diversity mean increased food and medicine resources. Food resources include plant and animal species that humans can sustainably harvest from highly diverse ecosystems as well as genetic traits that scientists can use in genetic modification of food crops grown elsewhere (see Chapter 11). Many medicines, such as antibiotics and heart, hypertension, and anticancer drugs, come from naturally occurring chemicals in plants.

✔ **Biodiversity builds more resilient ecosystems.** Every ecosystem experiences occasional disturbances, such as fires and floods. An ecosystem with high species diversity is more likely to bounce back from changes in the ecosystem. After all, with so many different species, a few of them are bound to be able to move in and rebuild the food web in a disturbed ecosystem.

✔ **Biodiversity supports ecosystem services.** *Ecosystem services* are the various natural functions, such as the decomposing of organic waste and the filtering and cleaning of freshwater, that humans benefit from. For example, as scientists now know, using a variety of wetland plant species is an effective way to clean wastewater (see Chapter 16).

✔ **Biodiversity is beautiful.** Some benefits of biodiversity are less concrete than medicine or waste decomposition. Biodiversity fuels a huge tourism industry that caters to people who value the beauty and recreation of a variety of ecosystems that are filled with a variety of species. The aesthetic value of biodiversity is more difficult to measure, but it's no less important when environmental scientists seek to preserve biodiversity (see the later section on conservation for details).

Becoming Biodiverse: Evolution by Natural Selection

The variety you see in the living things all around you is a result of each organism's unique genes. The *theory of evolution by natural selection* explains how species *evolve,* or change through time — sometimes changing so much that new species are created.

The theory of evolution by natural selection doesn't attempt to explain *how* life first began or *why* living things are on Earth. Rather, the theory of evolution by natural selection explains the scientifically observable processes that change the physical characteristics of living things through time.

Combining genetic material

Within each cell of an organism are numerous molecules forming deoxyribo-nucleic acid (DNA). In *eukaryotic,* or multi-celled, organisms such as humans, these molecules are called *chromosomes.* Each chromosome contains many *genes,* which are like the handbook for every complex system function of the cell as well as the physical characteristics of the organism. The physical characteristics that result from the genes are called *traits.*

When two organisms reproduce sexually, their chromosomes combine and create a new and unique sequence of information that determines the characteristics of the offspring. As a result, the offspring have some characteristics from one parent and some from the other.

A famous experiment by the scientist Gregor Mendel identified this reproduction process and predicted how the combination of genetic material allows parents to pass along traits. Without knowing anything about DNA or modern genetics, Mendel experimented with pea plants and tracked the physical traits (what are now known collectively as *gene expression*) in each generation. He recognized that each offspring receives two sets of information for every trait — one from each parent.

Mendel also discovered that some genes are dominant and others are recessive. A *dominant* gene is expressed as a physical trait when paired with a recessive gene. The *recessive* gene remains in the DNA code of the offspring, but it's physically invisible until it gets paired with another recessive gene in a later generation. This explanation of how genetic traits are passed from parent to offspring is called *Mendelian genetics,* or *classical genetics.*

Note: Although Mendel's explanation doesn't accurately describe every detail of what scientists now know about genetics and evolution, it does provide a simple and accurate description of what's occurring.

Selecting for survival

Each generation of offspring is the result of a novel combination of genes from each parent. However, sometimes random changes, called *mutations,* can occur by accident, or they may be caused by an outside influence. (I describe such influences, called *mutagens,* in Chapter 17.)

When a mutation occurs in a gene, it changes the gene and may change the physical trait determined by that gene. Sometimes change is good, sometimes it's bad, but most of the time it's neither good nor bad. Here's a quick look at the three types of traits that result from mutations:

- ✔ **Adaptive trait:** A new physical trait that in some way helps an organism survive is called an *adaptive trait.* For example, a trait that makes basic living (eating, growing, and reproducing) easier for an organism is an adaptive trait. As their name suggests, adaptive traits help organisms adapt to their environment, making living less difficult and, ultimately, reproduction more successful.

- ✔ **Maladaptive trait:** A new physical trait that disrupts an organism's life or decreases its chances for survival is called a *maladaptive trait.* Maladaptive traits interfere with an organism's ability to successfully reproduce and pass on its genes to the next generation. For example, a mutation leading to infertility is maladaptive because it interferes with an organism's reproductive success.

- ✔ **Neutral trait:** Some mutations are neither adaptive nor maladaptive; these mutations are called *neutral.* Neutral mutations sometimes lead to traits that don't affect the organism's ability to survive and reproduce in either a positive or negative way; hence, they're called *neutral traits.* *Note:* Sometimes neutral mutations don't express as any physical trait and, thus, don't affect the organism in any way whatsoever.

When the process of selecting for or choosing an adaptive trait occurs in nature, it's called *natural selection.* When humans manage this process, it's called *artificial selection.* Animal breeders, farmers, and gardeners play a role in evolution by selecting one trait over another. Unlike natural selection, artificial selection reduces the genetic diversity of the organism being worked with.

According to the theory of evolution by natural selection, mutations that result in adaptive traits are passed on to offspring (and selected naturally) because they assist an organism in surviving and reproducing. Eventually, all the changes may add up, creating a new species. The development of new species from an existing population is called *speciation*.

Scientists are still working to accurately define exactly what a species is. Right now, the most common and accepted definition is that a *species* is made up of organisms that are distinct from others in terms of size, shape, and behavior and that can breed with members of their own species to produce offspring. But even this definition has problems, since it doesn't apply to the many species of bacteria, which most commonly reproduce through asexual reproduction (and therefore don't breed with one another).

Another exception to this definition is dogs. For example, while all dogs are considered one species, different breeds of dogs are very different in size, shape, and behavior. In fact, these breeds are different enough that they don't interbreed. Yet their genetic information remains similar enough that scientists consider them all one species.

The theory of evolution by natural selection explains that living things change through time as a result of genetic mutations and natural selection for the most adaptive traits.

The HIPPO in the Room: Major Threats to Biodiversity

All around the globe you can find what environmental scientists call *biodiversity hotspots* — locations that have the highest biodiversity on Earth. Unfortunately, that high biodiversity is in danger of being reduced or damaged by human actions. Figure 12-1 illustrates where many of the biodiversity hotspots are located.

Multiple factors, usually related to humans, threaten biodiversity by causing species extinction. *Extinction* is when every single individual of a species has died. Scientists who study biodiversity have identified five major threats to biodiversity, or factors that lead to extinction. They summarize these threats with the acronym HIPPO:

H: Habitat destruction

I: Invasive species

P: Pollution

P: (Human) Population growth

O: Overharvesting

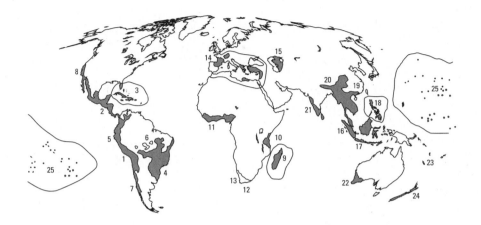

Figure 12-1:
Biodiversity
hotspots.

1. Tropical Andes, 2. Mesoamerica, 3. Caribbean, 4. Brazil's Atlantic Forests, 5. Chico/Darien/Western Equador, 6. Brazil's Cerrado, 7. Central Chile, 8. California Floristic Province, 9. Madagascar, 10. Eastern Arc & Coastal Forests of Tanzania/Kenya, 11. West African Forests, 12. Cape Floristic Province, 13. Succulent Karoo, 14. Mediterranean Basin, 15. Caucasus, 16. Sundland, 17. Wallacea, 18. Phillippines, 19. Indo-Burma, 20. South-Central China, 21. Western Ghats/Sri Lanka, 22. Southwest Australia, 23. New Caledonia, 24. New Zealand, 25. Polynesia/Micronesia.

Illustration by Wiley, Composition Services Graphics

Note: Environmental scientists today add *climate change* to the list of threats to biodiversity. As I describe in Chapter 19, predicted effects of the warming climate include shifting weather patterns and changing shorelines. The result could be further destruction of habitat and more opportunity for species to spread into new habitats as invasive species.

Each of these factors impacts biodiversity in a different way as a result of human actions, but the outcome is always the same: fewer species, less biodiversity, and weaker ecosystems.

Habitat destruction

Scientists recognize habitat destruction by human development and resource use as the top danger to biodiversity around the world. Anywhere humans build towns, villages, and cities, they destroy habitat. Unlike humans, most other species can't adapt to a wide variety of habitats. So after humans destroy the preferred habitat of certain species — by urbanization, land clearing for farmland, or some other human resource use — the species are left homeless and must either migrate to another location that suits their needs or perish.

Invasive species

An *invasive species* is a species that thrives in a new environment and out-competes the local species, disrupting the dynamics of the ecosystem (such as the food web or habitat structure). Invasive species are often *exotic, alien,* or *non-native species.*

Sometimes invasive species are brought to a new location by accident. For example, zebra mussels were transported in the ballast water of container ships and found in the Great Lakes in the 1980s. Since then, they have spread across freshwater habitats in the north central U.S., where they colonize and dominate ecosystems, clog pipes, jam boat motors, and even grow on the shells of other organisms. Other times humans introduce them on purpose. For example, humans brought kudzu to the southern U.S. to provide as a groundcover to control erosion. Now this woody vine covers nearly two million acres of land across the southern U.S., completely engulfing and crowding out other vegetation. More recently, legal action has been taken to consider pythons an invasive species. Originally brought to the U.S. as exotic pets, these large snakes have been released into ecosystems, such as the Florida everglades, where they kill a large number of birds.

The secret to success for many invasive species is that their new ecosystem doesn't contain the predators and resource limitations that controlled population growth in their original ecosystem. (See Chapter 8 for details on how resources and predators affect population size.) Upon entering the new ecosystem, some invasive species quickly adapt to their new resources and experience a population explosion. The following are some common invasive species in North America:

- **Cheat grass:** Also called *downy brome,* this grass crowds out native grasses in most of the western U.S.

- **Grass carp:** These large fish were brought to the U.S. to help control aquatic weeds. Since they were introduced, however, their populations have grown out of control and spread. Now grass carp are classified as invasive due to the damage they cause to aquatic vegetation in rivers.

- **Gypsy moth:** Originally brought to the U.S. as a silk producer, gypsy moths are capable of stripping the leaves from trees (especially oaks) to the point of destruction. Currently, they're found in the northeastern U.S., and quarantines are in place to control their spread.

- **Leafy spurge:** This invasive plant crowds out native species along the northern half of the U.S., from Michigan to the West Coast and as far south as Colorado.

- **Scotch broom:** This exotic ornamental shrub is invading the West and Southeastern coasts of the U.S.

- **Water hyacinth:** This aquatic plant species, which is found mostly in California, Texas, Louisiana, and Florida, grows dense colonies that block sunlight and crowd out native water plants.

Pollution

As I describe in Chapters 15 and 16, pollution has countless negative effects on the environment and human health. Similarly, pollution can lead to species extinction and a loss of biodiversity. When pollutants enter the environment — making water undrinkable or leaving dangerous chemicals in the air, water, and soil — every organism in the environment is affected. Some may react to pollutants more than others, and some may be exposed in higher doses (such as fish in areas of polluted water).

Due to the process of *biomagnification* (during which toxins become concentrated as they move through the food chain; see Chapter 17), animals at the top of the food chain are in more danger than others to the threat of pollution. These top predators often exist in smaller numbers, and as soon as pollution in the ecosystem begins to affect them, their populations quickly decrease.

Population growth (of the human variety)

When scientists look at threats to biodiversity, they recognize that human population growth underlies each of the others. The more people who live on the planet, the more habitat they destroy, the more invasive species they transport, the more pollution they create, and the more species they overharvest. Quite simply, to encourage (rather than destroy) biodiversity around the world, growing human populations need to think about sustainability now and into the future.

Overharvesting

You may not realize that *overharvesting*, or consuming more plants and animals than a population can replace through reproduction, threatens many species. Humans overharvest some species, such as fish, as food resources and others, such as saltwater aquarium fish and orchids, for collectors.

Another use that leads to overharvesting is the medicinal or folk use of rare species. For example, humans kill black rhinos for their horn, which they then sell as a medicinal cure for a variety of symptoms in Eastern Asia. Similarly, many plants, such as ginseng, are overharvested to meet consumer demand for their medicinal properties.

Creating Effective Conservation Plans

Recognizing the importance and vulnerability of ecosystems, many nations have created conservation plans to preserve biodiversity and policies to encourage both sustainability and biodiversity preservation. Some conservation laws protect endangered or threatened species, and others help manage harvesting in more sustainable ways.

One example is the Endangered Species Act (ESA) in the U.S. The ESA identifies species at risk for extinction, develops plans to recover or maintain their populations, helps landowners implement these plans, and enforces the protection of species and habitat. The ESA defines three levels of extinction risk:

- **Endangered species:** *Endangered species* are considered to be in the most danger of becoming extinct.

- **Threatened species:** *Threatened species* are likely to become endangered in the near future.

- **Vulnerable species:** *Vulnerable species* are rare species, whose populations have been depleted to a great degree because of human actions. These species may eventually become endangered.

The International Union for Conservation of Nature and Natural Resources (IUCN) categorizes species along similar lines. But unlike the U.S., the IUCN doesn't create recovery plans or enforce any conservation laws.

In countries like the U.S. that do create and implement recovery plans, conservation scientists and policymakers have developed two very different approaches, which I describe in the next sections.

Of singular importance: The species approach to conservation

Single-species conservation plans, like the Environmental Protection Act in the U.S., look at each individual species separately and focus on improving environmental conditions for a single species by reducing pollution and habitat destruction in an effort to boost population growth for that species. So while the conservation or recovery plan focuses on a single species, the entire ecosystem benefits from a cleaner, healthier environment and reduced human impacts.

So how does a specific species become the focus of a conservation plan? Most conservation plans single out certain species because they play an important role in the ecosystem. Scientists recognize the following special species:

- **Keystone species:** A *keystone species* is one that plays a particularly important role in its ecosystem. The extinction of a keystone species would affect many other organisms in the ecosystem. For example, scientists have recognized wolves as a keystone species in some regions, where they help control deer and moose populations.

- **Indicator species:** An *indicator species* is one that indicates, or represents, an aspect of the environment or ecosystem in which it lives because it is specially adapted or limited to that particular aspect. Indicator species are very sensitive to changes in their environment. When these species aren't thriving, something in the environment may be out of balance; for example, it may be experiencing too much pollution. Certain species of coral act as indicator species in coral reef ecosystems because they die if the pH or salinity (amount of dissolved salt) of the water changes.

- **Umbrella species:** An *umbrella species* is one that requires a large amount of habitat. When conservation plans focus on umbrella-species conservation, they protect large regions of habitat, and this protection benefits every species living there. An example of an umbrella species is the northern spotted owl in the Pacific Northwest of the U.S.

- **Flagship species:** *Flagship species* are the cuddly and attractive animals or especially beautiful plants that people feel emotionally attached to. Environmental scientists often use these species to promote conservation issues and encourage people to become involved and play a role in preserving biodiversity. The most common flagship species are the giant panda bear, the California redwood, and more recently, the polar bear.

In the species-based approach to conservation, scientists may breed individuals of an endangered species in captivity in order to preserve and restore their populations. But what good is a growing population of mountain gorillas in the zoo if they have no mountains to return to after their populations have recovered? Scientists have answered this question by presenting an ecosystem-based or habitat-centered conservation approach.

Size does matter: Preserving entire ecosystems

Another approach to conservation — called *ecosystem-based* or *habitat-centered conservation* — looks beyond single species and seeks to protect entire ecosystems. Preserving a whole habitat from destruction protects all the species that live there and maintains ecosystems for species that are recovering from near extinction.

However, simply fencing off areas of high biodiversity isn't enough. Conservation scientists have realized in the last few decades that each preserved space creates an island of protected habitat but leaves the edges and the spaces in between unprotected. To tackle this issue, environmental scientists have explored different designs for habitat conservation that help extend protected habitat and reduce the effects of human populations around the edges of conserved areas.

Scientists look at how best to protect biodiversity through habitat conservation in two different ways. One is the island biogeography theory, and the other is through the creation of biosphere reserves.

Focusing on islands in the landscape

The *theory of island biogeography* explains how the size of a habitat affects the biodiversity of an ecosystem. Originally used to study actual island ecosystems, such as the Galapagos Islands and Hawaii, scientists now apply island biogeography principles to "islands" of protected or undisturbed landscape.

Island biogeography studies have shown two key points that relate specifically to biodiversity conservation:

- In general, larger habitats are home to a larger number of species and larger populations of each species.

✔ In some cases, the overall size of a preserve isn't as important as whether multiple protected areas are connected together. When multiple areas are connected, species have the opportunity to migrate over greater distances and interact with other populations, thus maintaining species richness across the ecosystem.

In regions where multiple islands of preserved habitat already exist, conservation scientists propose connecting the islands together with *protected corridors,* or narrow regions of protected habitat linking one island to another. Connecting areas of landscape with protected corridors, such as the one illustrated in Figure 12-2, allows species to travel between the islands and greatly increases their habitat size. These corridors also allow humans and other species to inhabit a region together.

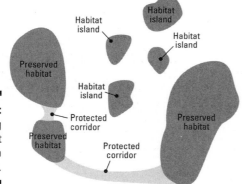

Figure 12-2:
Connecting habitat islands with corridors.

Illustration by Wiley, Composition Services Graphics

Buffering the core

Another approach to conservation that tries to balance human resource use with healthy and protected ecosystems is the creation of *biosphere reserves.* Biosphere reserves are unique because they're designed to reach beyond habitat preservation or resource conservation. The three main goals of a productive biosphere reserve are

✔ Conservation of biodiversity

✔ Research and monitoring of ecosystem health

✔ Promotion of sustainable development

To achieve these goals, most biosphere reserves are divided into three separate zones, as illustrated in Figure 12-3:

- ✔ **Core area:** Each biosphere reserve has at its center a *core area* that's protected from resource harvesting or development. Research scientists may monitor the core area to measure the overall health and biodiversity of the ecosystem.

- ✔ **Buffer zone:** Surrounding the core area is the *buffer zone,* a region where scientists and others practice research, education, and ecological tourism with the goal of encouraging people to value the habitat conservation at the core. (Check out the nearby sidebar "Touring sustainable ecosystems" for details on sustainable tourism.)

- ✔ **Transition zone:** Beyond the buffer zone is the *transition zone,* where you find sustainable development and human settlements. In this zone, humans may harvest resources by using sustainable practices.

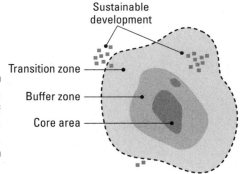

Figure 12-3: The zones of a biosphere reserve.

Illustration by Wiley, Composition Services Graphics

By establishing this sequence of zones, a biosphere reserve protects the core area, including its edges, from unsustainable resource harvesting and environmental damage. At the same time, humans can meet their resource needs, and local populations get educated in sustainable practices to extend habitat and resource conservation into the future.

Touring sustainable ecosystems

Ecological tourism or *ecotourism* is an effort to create sustainable economies in biologically diverse and vulnerable ecosystems of the world. Ecotourism attracts visitors who are interested in viewing preserved landscapes in a way that has very little impact on the landscapes themselves. Often on a much smaller scale than other forms of tourism, ecotourism invites tourists to travel into the educational and research zones surrounding a protected core of wilderness in the hopes that the visit will inspire those tourists to value wilderness, biodiversity, and sustainability.

The financial profit from ecotourism industries goes to support sustainable local economies and promote further low-impact use of the protected areas for education and advocacy. Another hallmark of ecotourism is that it respects and protects traditional ways of life in parts of the world that aren't fully urbanized. Working together with local populations in vulnerable and biodiverse regions provides native cultures with an opportunity to play a pivotal role in preserving the landscape and ecosystems they depend on.

Here are a few tips in case you're interested in becoming an ecotourist:

- ✔ Study up on the history and culture of the area you'll be visiting before you go.

- ✔ While visiting protected areas, take only pictures and leave only footprints.

- ✔ Be mindful of your personal resource use by conserving water, food, and fuel and by considering where your waste and garbage are disposed.

- ✔ Be considerate of cultural differences and respect sites of cultural and religious importance to the local people.

- ✔ Don't purchase items that may have been illegally or unsustainably harvested.

- ✔ Consider sharing your ecotourism experience with others through a journal, blog, or informal presentation when you arrive home. Education is the key to maintaining interest in preserving worldwide biodiversity!

Chapter 13

Hitting the Hard Stuff: Geologic Resources and Energy

Despite being mostly covered by water, the surface of the Earth is a rocky place. These rocks hide many important and useful resources — in particular metals, minerals, and fossil fuels. Removing geologic resources that are valuable for industry (such as metals) or for energy (such as fossil fuels) can easily result in environmental damage and endanger human health. Environmental scientists seek sustainable ways to extract these resources while protecting the ecosystems and preserving the landscapes they're hidden within.

In this chapter, I introduce a few basic concepts in geology, such as the rock cycle and plate tectonics. I explain how economically useful rocks and minerals are mined from the Earth, and I cover the environmental effects of mining activity. In the final sections of the chapter, I focus on fossil fuels and nuclear energy. Fossil fuels such as coal, oil, and natural gas are the most commonly used energy source for industry and transportation today. I explain why the supply of fossil fuels is limited and describe the pros and cons of nuclear energy as an alternative.

Getting to Know the Earth

Geology is the study of the Earth. Geologists observe and draw conclusions about the features of Earth's surface and the processes that create them.

Some of geology's big ideas are useful to know as you study environmental science; here are some facts to keep in mind:

- **The Earth is composed of multiple layers.** The innermost layer of the Earth is a solid metal *inner core.* Surrounding the solid core is a liquid *outer core* of metal. Beyond the outer core is the *mantle,* a thick layer of solid and flowing rock material. Atop the mantle is the Earth's *crust,* the solid layer of rock that you're most familiar with.

- **The Earth's crust is composed of multiple plates.** The surface of the Earth is broken into pieces, called *tectonic plates,* that move around in response to energy transferred from the mantle through the process of heat convection (see Chapter 4). *Plate tectonics theory* explains how the interaction of tectonic plates creates geologic features such as volcanoes and mountains.

- **The Earth's crust is composed primarily of eight elements.** The most common elements in Earth's outer layer are oxygen, silicon, aluminum, iron, calcium, magnesium, sodium, and potassium. These elements combine to form *minerals* (naturally occurring, solid compounds with defined compositions and crystal shapes), which combine to form *rocks.*

- **Rocks are formed, transformed, and recycled through the rock cycle.** Processes of plate tectonics and exposure to the atmosphere change rocks from one type into another. This cycle has no beginning or end; any rock at any stage of the rock cycle can be transformed into a different rock if exposed to the right conditions. Figure 13-1 illustrates the basic processes and rock types of the rock cycle.

Thinking geologically about environmental issues

The fields of geology and environmental science overlap in the study and practice of *environmental geology*. Environmental geologists are trained in geology and play an important role in managing geologic resources (fossil fuels and minerals, as well as groundwater) and surface resources (such as water in streams and rivers). They have a deep understanding of how environmental damage and pollutants can affect the sediments, rock, and water on and below Earth's surface.

Many environmental geologists are engineers who use their geological knowledge to manage geological resource extraction in safe and cost-effective ways. Others apply their engineering skills to solve environmental problems such as stabilizing abandoned mine sites, protecting groundwater from pollutants, and restoring natural hydrology (water flow) to reduce erosion.

Environmental geologists also apply their training to helping communities prepare for and minimize the damage from natural disasters. Earthquakes, volcanic eruptions, floods, and tsunamis dramatically affect ecosystems and human communities. By understanding the likely outcome of such events, environmental geologists help communities asses their potential risk, create evacuation plans, and restore what has already been damaged.

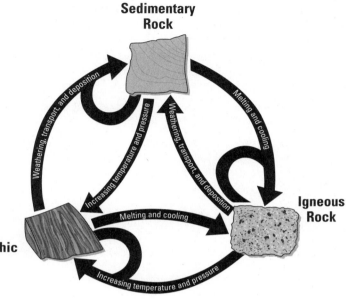

Sedimentary Rock

Weathering, transport, and deposition

Melting and cooling

Weathering, transport, and deposition

Increasing temperature and pressure

Melting and cooling

Igneous Rock

Metamorphic Rock

Increasing temperature and pressure

Figure 13-1:
The rock cycle.

Illustration by Wiley, Composition Services Graphics

Geologic resources that are valuable for fuel, construction, and manufacturing result from processes of plate tectonics and the rock cycle. I describe some of these resources in the next section.

Finding Value in Rocks

Important geologic resources include rocks found in the Earth's crust and the minerals they contain. Some minerals are extremely valuable because they hold important elements (such as uranium) or metals (such as gold). Other minerals, such as salts, are less valuable and more abundant. In this section, I describe some of the most commonly used and economically valuable geologic resources.

Rocks and minerals

Some of the most abundant geologic resources are rocks composed of common minerals such as the following:

- **Silica:** *Silica* is a very common mineral made up of silicon and oxygen (the two most common elements in Earth's crust). Silica is used to make glass, and most sand is made of silica minerals.

- **Calcium carbonate (limestone):** *Limestone* is a sedimentary rock that is a common ingredient in cement, concrete, and (in its crushed form) road gravel.

- **Halite (salt):** *Halite* is the mineral that provides table salt. In its raw form, road crews also use it to melt ice on roads in the winter.

- **Gypsum:** *Gypsum* minerals are used to make plaster and plasterboard for buildings. Gypsum is the white, ash-like stuff that comes out of your wall if you drill a hole in it to hang a painting.

Metals and ores

The Earth's most valuable mineral resources aren't the most common ones. The most economically important are *metals,* especially ones that are relatively rare in Earth's crust. Metals in the crust are usually a smaller portion of a larger mineral called an *ore* and require extensive processing to be purified and concentrated into the desired metal.

Although some metals, such as gold and silver, are used as currency, the real value of these and other metals is in their industrial uses. Table 13-1 lists the common uses of metals in the U.S.

Table 13-1	Common Uses of Metal Resources
Metal	*Use*
Aluminum	Packaging and electronics
Copper	Construction, electrical wires
Lead	Paint, gasoline, ammunition
Iron	Steel production, machinery
Nickel	Chemical production, steel production
Gold	Electronics, currency
Silver	Photography, electronics, jewelry, currency

The ores containing these (and other) metals are often found in mountainous regions of the world, where people mine them in ways that cause extensive damage to the surrounding ecosystem. (I provide details on mining in the next section.) The worldwide distribution of metal resources plays an important role in determining which nations benefit from the production, export, and use of metals. For example, the U.S., China, Russia, and Australia have large reserves of metal ores, while the U.S., Japan, and European nations consume the most metal resources. In some places where metals are rare, such as Africa, access to the limited supplies fuels civil wars and leads to environmental exploitation.

In Chapter 18, I describe how electronic waste is often sent to African or Asian countries, where the poorest people expose themselves to serious toxins in order to scavenge bits of valuable metals, such as copper wiring, from the trash. Although this practice is dangerous for human health, it illustrates that many metals used in industry and commercial products can be recycled by manufacturers or through scrap metal collectors.

In North America, regulations require new steel products to be composed of 28 percent recycled metals. To meet these requirements, the steel industry has developed new methods of steel manufacture that reuse and recycle scrap iron and steel. Small-scale steel mills called *minimills* are an efficient way to recycle scrap steel and iron into new products. These mills are smaller than traditional steel mills yet still capable of processing metals from ores to finished products. Some minimills create new steel products with up to 90 percent recycled metals!

Extracting Geologic Resources

Because the most valuable mineral resources are buried deep within Earth's crust, they must be *mined,* or extracted, before they're usable. Most mining methods result in some environmental damage because humans are removing a resource from the ecosystem for their own use. Unfortunately, many mining operations result in air and water pollution as well. In this section, I describe the most common methods of surface and subsurface mining, including the environmental effects of each technique.

Scraping the surface: Surface mining

Valuable geologic resources often occur at or near the Earth's surface. To access these materials for extraction, miners rely on a few commonly used methods of *surface mining*. Humans often extract resources such as copper and coal by using massive surface mining operations.

The environmental damage caused by surface mining is related to the large amount of surface material that humans remove during mining operations. The environmental effects of surface mining include

- Habitat destruction
- Soil erosion
- Air pollution from dust particulates
- Pollution (especially from sediments)

All surface mining techniques negatively affect the environment, though some methods are more damaging than others, as I describe in the following sections.

Strip mining

Strip mining, as the name describes, is a process of removing rock and soil in strips to get to the valuable mineral ores below. After miners extract the resources, they put back the leftover rock and soil, called *mining spoils* or *tailings,* to fill in the hole. One way to replace tailings is to simply dump them; luckily, U.S. regulations require mining companies to replace tailings in a way that restores the landscape (and ecosystem) more closely to its pre-mined state, even though doing so is often expensive and difficult. When tailings aren't restored properly, they're often left in valleys where they cause flooding and disrupt watershed ecosystems with water pollution and increased sediments.

Mountaintop removal mining

Mountaintop removal mining is similar to the strip mining approach in the preceding section but on a much larger scale. This technique removes large amounts of rock and soil — whole mountaintops — to access the resources buried deep inside the mountain. The mountaintop material is left in surrounding areas of lower elevation, such as nearby river and stream valleys, where it reshapes the landscape, pollutes water, and disrupts ecosystems. Figure 13-2 illustrates how mountaintop removal mining dramatically changes a mountainous landscape.

Figure 13-2:
Mountain-
top removal
mining.

Photograph by Getty Images/Pete Mcbride

Pit mining

Sometimes valuable geologic resources appear at the surface of the Earth but extend deeper into the Earth — sort of a tip-of-the-iceberg kind of thing. In this situation, *pit mining* (also called *open-pit mining*) is an option. Pit mining involves digging a large hole to gather rocks and minerals from the Earth's crust. Pit mines extend both into the ground and across the surface and are some of the largest mine operations in the world. Like other surface mining operations that remove materials, pit mining operations scar the landscape, destroy habitat, and pollute the air with dust and particulates.

Placer mining

Placer mining is a way of obtaining mineral and metal resources from loose river sediments. The water helps sift the valuable resources (such as gemstones or gold) from the sand, mud, and gravel in the riverbed. Placer mining occurs on a much smaller scale than other methods of surface mining. Whereas other methods go straight to the source of the valuable material, the sought-after materials in placer mining have already been removed from their source by natural processes of erosion and weathering and have been carried downstream. Due to the smaller scale of placer mining, it does less damage to the surrounding environment compared to other methods of surface mining, though it can still disrupt river ecosystems with pollution and sediments.

In some places, such as the Yukon Territory in Canada where placer mining is common, miners make an effort to maintain the water quality by periodically testing for pollution and sediment overload. In some areas, miners have even developed placer mining systems that recycle the water used to sift gold so that polluted water isn't discharged back into the environment.

Digging deep: Subsurface mining

Surface mining techniques don't work for extracting all valuable geologic resources. Diamonds and most metal ores, including gold, require extensive subsurface mines to access the rocks with these resources in them. *Subsurface mines* are probably what you envision when you think of mining: systems of tunnels and vertical shafts with elevators to take miners underground where they can retrieve the valuable resources.

Subsurface mining operations don't create the visible changes in the landscape that surface mining does, but the conditions of subsurface mines are extremely hazardous for the working miners. The potential for accidental cave-ins, explosions, and fires is high. The air quality deep within the mines is poor; the atmosphere is rife with particulates and gases that lead to respiratory diseases, including lung cancer.

Subsurface mines also produce large amounts of environmentally hazardous *acid mine drainage*. To keep the underground system of tunnels and mine shafts clear, mining companies have to pump out large amounts of water, which go into surface ecosystems. The groundwater from the mines is more acidic than surface waters and disrupts ecosystems by changing the pH conditions of soil and water sources. (Check out Chapter 3 for more on pH.)

Some mining companies have begun to realize that keeping mining operations environmentally safe and clean from the start is more cost-effective. Mining corporations prefer to avoid the expense of cleanup and restoration or of being held legally responsible for ecosystem destruction or human health effects. Instead, these companies see that some investment into sustainable mining practices saves them money in the long run.

Forming and Depleting Fossil Fuels

Fossil fuels, including coal, oil, and natural gas, generate almost all the world's energy. In this section, I explain how the rock cycle forms these important and nonrenewable geologic resources over millions of years. Then I describe how fossil fuels are extracted, and I touch on the pros and cons of depending on them for fuel.

Creating rocks from life

The secret to fossil fuels' ability to produce energy is that they contain a large amount of carbon. This carbon is leftover from living matter — primarily plants — that lived millions of years ago. Oil and natural gas are usually the result of lots of biological matter that settles to the seafloor, where the *hydrocarbons* (molecules of hydrogen and carbon), including methane gas, become trapped in rocks.

Coal's formation is a little different. Coal starts as *peat,* or sections of partially decomposed organic matter that accumulate on the Earth's surface. Over millions of years, the peat is buried and heat and pressure transform it into increasingly pure forms of coal called *lignite, sub-bituminous, bituminous,* and *anthracite coal.*

These fossil fuels are limited or *nonrenewable* resources; after they serve their fuel purpose, they can't be recycled back into a useful energy source. Their supplies are *finite;* when they're gone, they're gone. True, more coal, oil, and gas supplies may be created over time, but that won't occur in our lifetime; as I note earlier, fossil fuel production takes many millions of years. (Don't despair; in Chapter 14, I describe renewable sources of energy as an alternative to fossil fuels.)

Down and dirty: Mining and burning coal

Coal is found in layers of rock that have been compacted and folded into mountains. Coal resources are fairly abundant throughout the world, though like any geologic resource, they aren't evenly distributed. The largest coal sources in the world appear in the U.S., Russia, China, India, and Australia. The purest coal forms (bituminous and anthracite) provide the most energy, but in general, coal requires little to no refining before it can be burned as fuel. Coal's abundance and ease of use make it an inexpensive fuel resource, particularly for developing nations that don't yet have fancy industrial refineries.

Using coal for fuel has several downsides, including the following:

- ✔ **Burning coal releases toxins.** Coal contains sulfur and other elements, including dangerous metals such as mercury, lead, and arsenic, that escape into the air when coal is burned. Burning coal also produces particulates that increase air pollution and health dangers (see Chapter 15 for more details, including how sulfur in the air creates acid rain).

- ✔ **Burning coal emits large amounts of carbon dioxide into the atmosphere.** Coal is composed almost entirely of carbon, so burning coal unleashes large amounts of carbon dioxide (CO_2) into the atmosphere. These emissions have been shown to increase the greenhouse effect in the atmosphere and lead to global warming. (Flip to Chapter 19 for details on global warming.)

- ✔ **Subsurface coal mining is dangerous.** Coal is often mined in subsurface mines, which may collapse and trap miners. And the air in subsurface coal mines leads to *black lung disease,* where coal particles and pollutants fill the lungs and cause inflammation and respiratory illness. For more info on this kind of mining, head to the earlier section "Digging deep: Subsurface mining."

- ✔ **Surface coal mining damages the environment.** Mountaintop removal mining is used to access layers of coal buried deep within mountains. As I describe earlier in the chapter (see the section "Mountaintop removal mining"), this mining technique alters the landscape and damages ecosystems.

Because coal is so abundant and relatively inexpensive, many people are reluctant to give it up as a fuel source. Luckily, ways to use coal more sustainably and minimize its environmental damage are available. Clean coal solutions include the following:

- ✔ **Integrated gasification combined cycle (IGCC):** *IGCC* technology converts coal into gas, removing sulfur and metals. This gas generates electricity by fueling turbines while the side products (sulfur and metals) are concentrated and sold. IGCC plants are cleaner and more efficient than coal-burning electric plants and have the potential to capture CO_2 emissions in the future.

✔ **Carbon sequestration:** One of the biggest problems with burning coal is the amount of CO_2 it adds to the atmosphere. *Carbon sequestration* includes various ways to capture and store carbon underground instead of allowing it to fill the atmosphere. Currently, some coal-burning plants store carbon in underground abandoned mines or in oil wells. Other plants pump the carbon into sedimentary rocks or below the ocean floor.

Perusing petroleum and natural gas resources

What you know as oil is actually called *petroleum* or *crude oil* and may exist as a combination of liquid, gas, and sticky, tar-like substances. Petroleum sources are usually small pockets of liquid or gas trapped within rock layers deep underground (often under the seafloor). Extracted crude oil is refined and used to manufacture gasoline (used in transportation) and *petrochemicals* (used in the production of plastics, pharmaceuticals, and cleaning products).

Like other resources, oil isn't evenly distributed across the globe. The top oil-producing countries are Saudi Arabia, Russia, the U.S., Iran, China, Canada, and Mexico. Together, these countries produce more than half of the total oil resources in the world.

While some petroleum is found in gas form, the most common *natural gas* is methane. Methane usually occurs in small amounts with petroleum deposits and is often extracted at the same time as the petroleum. Natural gas can be found in certain rock layers, trapped in the tiny spaces in sedimentary rocks. The following sections give you the lowdown on these fuels and introduce you to a couple of lesser-known petroleum-related resources.

Drilling for oil

Oil companies pump liquid oil out of the ground by using *drilling rigs* and wells that access the pockets of oil resources. The oil fills the rock layers the way water fills a sponge — spreading throughout open spaces — instead of existing as a giant pool of liquid.

This arrangement means that to pump out all the oil, drillers have to extend or relocate the wells after the immediate area has been emptied. Oil drilling rigs set on platforms in the ocean to access oil reserves below the seafloor must therefore employ a series of more technically complex drill rigs built to access oil reserves in deeper water. Figure 13-3 illustrates some of the most commonly used ocean drilling rigs and platforms and the water depths they're most suited for.

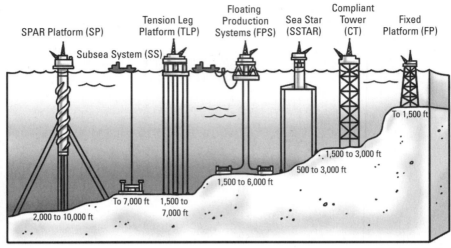

Figure 13-3: Seafloor oil drilling rigs.

Illustration by Lisa Reed

Oil is a cleaner fuel than coal, but it still has many disadvantages, such as the following:

- **Refining petroleum creates air pollution.** Transforming crude oil into petrochemicals releases toxins into the atmosphere that are dangerous for human and ecosystem health.

- **Burning gasoline releases CO_2.** Although oil doesn't produce the same amount of CO_2 that coal burning does, it still contributes greenhouse gases to the atmosphere and increases global warming.

- **Oil spills cause great environmental damage.** Large oil spills sometimes occur during drilling, transport, and use, which of course affects the surrounding environment. But these spills aren't the only risk. Although large oil spills with catastrophic environmental effects — such as the 1989 *Exxon Valdez* in Alaska or the 2010 BP *Deepwater Horizon* in the Gulf of Mexico — get the most media coverage, most of the oil spilled into ecosystems is actually from oil that leaks from cars, airplanes, and boats, as well as illegal dumping.

Fracking for natural gas

Natural gas is a relatively clean-burning fuel source — it produces approximately half the CO_2 emissions that coal burning produces — so demand for natural gas has increased in the last few decades as concerns grow about carbon emissions and global warming. Now fuel producers are exploring natural gas in reservoirs separate from petroleum as sources of this fuel. To release the gas from the rocks and capture it for use as fuel, companies use a method of *hydraulic fracturing,* or *fracking.*

Fracking for natural gas requires injecting a liquid mix of chemicals, sand, and water into the gas-bearing rock at super high pressures — high enough to crack open the rock, releasing trapped gases. The gas is then pumped out of the rock along with the contaminated water. The sand and chemicals are left behind in the rock fractures, leading to groundwater pollution and potentially less stable bedrock. Currently scientists are concerned that earthquakes in regions of the Midwestern U.S. that have never experienced earthquakes before are the result of wastewater from natural gas fracking operations.

Unconventional petroleum resources

Although oil and natural gas are the most common petroleum resources, other similar, lesser-known resources are available:

- **Tar sands:** In some parts of the world (such as Canada and Venezuela), large deposits of sand are mixed with tar or *bitumen,* a sticky hydro-carbon substance. Although the tar sand resources are vast in these regions, they have high environmental costs, such as the habitat destruction required to extract them and the production of greenhouse gases and toxic waste in the refinery process.

- **Oil shales:** *Oil shales* are sedimentary rocks that contain *kerogen,* an oil-like substance. The current process for extracting the oil from these rocks involves using and polluting large amounts of water. So far, researchers haven't found an environmentally safe and economically reasonable way to access these fossil fuel resources, but research continues.

Reacting to Nuclear Energy

Producing nuclear energy relies on supplies of uranium, a geologic resource. The development of modern nuclear reactors has slowed in the last few decades because fears about potential accidents have outweighed the benefits of nuclear power. In this section, I briefly explain how nuclear power is generated and describe the pros and cons of using nuclear energy.

Splitting atoms: Nuclear fission

A process of atomic fission creates nuclear energy. *Fission* is, in simplest terms, the splitting of an atom. To generate nuclear energy, uranium atoms are split into two separate atoms. This reaction is *exothermic,* which means it produces heat energy as the atom splits in two. The energy released from the fission reaction sets off a chain reaction, leading more uranium atoms to split and release more energy. This powerful sequence of exothermic reactions produces a huge amount of heat energy, which must be carefully managed and captured for use.

Uranium used in nuclear power plants is packed into highly concentrated pellets and bundled into fuel rods, which are located in the reactor core of the nuclear power plant. Controlling the highly reactive situation that results from uranium fission's chain reaction requires a complex cooling system. The nuclear reactor cooling system circulates water through the fuel rods in the core, keeping temperatures cool enough to avoid a nuclear meltdown.

Fusing atoms: Nuclear fusion

Another way to generate nuclear energy is through *nuclear fusion,* the process of combining atoms in a reaction that produces energy. This process is how the sun generates the heat and radiation that warm and light the Earth. Scientists understand how to generate nuclear fusion, but they don't yet have a way to safely create the special conditions of heat — such as those found in the core of the sun — necessary to achieve fusion. Until they do, nuclear fusion isn't an energy option.

The once and future fuel: Debating nuclear energy

As environmental scientists and the industrial world look for ways to reduce the dependence on fossil fuels, nuclear energy seems to provide a potential solution. As with any fuel source, though, using nuclear energy has pros and cons.

One of the biggest advantages in using nuclear power (especially compared to some of the alternative energy sources I describe in Chapter 14) is the ability to steadily produce large amounts of energy. Nuclear power produces high amounts of energy at a steady rate, so it can address many of the energy needs currently being met by coal. Additionally, shifting some energy dependence away from fossil fuels and onto nuclear energy may reduce international tensions as petroleum resources shrink.

The cons of using nuclear energy include the following:

- ✔ **Finding safe long-term storage for waste is vital.** Nuclear energy generation results in radioactive waste that needs safe, secure, and long-term storage.

- ✔ **Uranium mining causes environmental damage.** Like many mining activities, extracting uranium ores leaves waste material at the site and requires large amounts of fossil fuels to operate mining machinery and transport uranium for processing.

✔ **People fear nuclear accidents.** Nuclear meltdowns are many people's biggest fear about nuclear energy. Nuclear power plants don't explode like nuclear bombs do; however, if the plant's cooling systems aren't functioning properly, the reactor core becomes too hot and begins to release the radioactive byproducts of nuclear fission into the atmosphere, water, and soil.

In 1986, a nuclear reactor in Chernobyl, Ukraine, experienced a meltdown while testing its cooling system (which failed). The result was the release of huge amounts of radioactive material into the atmosphere; additionally, portions of the plant caught fire and unleashed contaminated smoke. Much of Europe and the Soviet Union were affected by the radiation because weather patterns spread the emissions across the continent. Regions in and around Chernobyl are still mostly toxic and uninhabited.

In March 2011, following an earthquake and tsunami in Japan, the Fukushima Daiichi nuclear power plant experienced a cooling system failure and a meltdown that released radioactive material into the ocean and atmosphere. Considered the largest nuclear accident since Chernobyl, the effects of this event are still being studied.

Many environmental scientists see nuclear energy as a useful tool to transition from fossil fuels to renewable energy in the future. However, research and technology is necessary to provide the safest, cleanest, and most efficient methods of capturing energy from nuclear reactions.

Chapter 14

Shifting Gears: Alternative Energy

· ·

In This Chapter

▶ Using organic matter for fuel

▶ Generating power with water and wind

▶ Finding heat energy deep in the Earth and converting sunlight to power

▶ Developing fuel cell technology

▶ Understanding the importance of energy conservation

· ·

*M*odern living requires huge amounts of energy. From the lights you turn on in your house to the goods you consume each day, everything requires an energy source. Where does all this fuel come from? In the past, most of it came from fossil fuels, but over the last few decades, as scientists have come to understand the role of fossil fuels in the changing climate, they've begun to pursue more research and development in alternative energy sources. In fact, one of the top concerns for environmental scientists these days is finding ways to harness energy from renewable sources and use that energy in the most efficient way possible.

In this chapter, I describe the current alternatives to fossil fuel energy, including the technology that captures energy from rivers, waves, sunlight, and wind. I also describe developing technologies that use methane gas, corn, algae, and other biological matter as fuel sources. I end the chapter by explaining how energy efficiency and conservation are energy sources of their own and telling you how you can get in on the action in your own home.

Looking for Alternative Energy

Most of the energy fueling today's industrial societies and the developing world comes from burning fossil fuels, such as coal, oil, and natural gas. Fossil fuel and nuclear energy sources rely on finding and then destroying geologic resources, such as coal, plutonium, and uranium. These resources are in limited supply on the planet, and over the last century, coal, oil, and natural gas in particular have become more difficult to access. At some point, the cost of extracting and processing fossil fuels will be more than the returns in energy they provide. And while nuclear energy is a powerful alternative, it also relies on stores of the geologic resource uranium, which is in abundant but still limited supplies (see Chapter 13 for more details on fossil fuels and nuclear power).

Alternative or renewable energy sources aren't based on resources that are in limited supply on Earth; instead, they're captured from processes (such as wind, waves, and sunshine) that are continually being driven by energy from the sun or created using materials (such as water) that are naturally renewed through Earth's processes. Table 14-1 lists some alternative energy resources and their associated pros and cons, as well as any pollutants they create.

Table 14-1	Alternative Energy Resources		
Energy Source	*Pro*	*Con*	*Emissions/Pollutants*
Biofuel	Easily grown anywhere	Competes with food crops	Particulates (small particles of solid or liquid material suspended in the air; see Chapter 15)
Water	Endlessly renewable, clean	Some habitat destruction	None
Geothermal	Low cost after installation	Geographically limited	None
Wind	Low cost to install	Geographically limited	Potential danger to flying organisms (birds, bats, and butterflies)
Solar	Endlessly renewable	High initial costs	None
Fuel cells	Very efficient, no pollution	Difficult storage and transport	None (if fossil fuels aren't needed to process and transport them)

Farming Fuel: Biofuel Energy Sources

Sources of energy that come from organic matter or organic processes are called *biofuels*. Biofuels come in several different forms, but what they all have in common is that they have captured energy from sunlight and stored it as organic matter, such as plant tissue. In a sense, fossil fuels like coal, which I cover in Chapter 13, can be considered biofuels because they're the fossil remains of ancient organic materials. In this section, I describe non-fossil fuels, or *modern carbon fuels,* that are considered renewable resources.

Biomass

The term *biomass* describes plant material or animal waste that can be burned as fuel. The best part about biomass is that you can grow it almost anywhere. The most common biomass fuel is wood. In many parts of the world, firewood is the primary fuel source that families use to meet their daily needs like cooking and heating. Elsewhere, people use animal waste, or *dung,* as a biomass fuel.

Using wood as a fuel source is only renewable if humans don't harvest the wood more quickly than it can be regrown. Some trees, shrubs, and plants grow quickly, so humans can plant them in large plantations to provide a renewable source of biomass fuel.

Some people suggest a downside to burning biomass for energy is that it releases carbon dioxide (CO_2) into the atmosphere, which affects the global climate system (see Chapter 19). However, this problem is easy to correct by replanting.

Because the processes of photosynthesis and respiration balance each other out, one solution to the CO_2 problem is to sustainably manage biomass fields so that the planting of new trees (which absorb CO_2) balances out the CO_2 emitted from burning harvested trees.

Biodiesel, ethanol, and other liquid biofuels

Liquid biofuels, such as biodiesel and ethanol, have the potential to replace fossil fuels as energy sources for transportation and heating. After all, you can grow the raw materials (plants) for liquid biofuels anywhere, unlike fossil fuel sources, which are unevenly distributed, are controlled by certain nations, and require long-distance transportation to get to where they're needed.

Biodiesel is made from vegetable oils much like the ones you use to cook with in your kitchen. It can replace petroleum gasoline to fuel vehicles and other fossil fuels as a heating oil in furnaces. Most biodiesel today comes from soybeans, but scientists have recently been working on ways to produce it from algae. *Algae,* a single-celled microscopic organism that uses photosynthesis to capture energy from sunlight, offers many benefits as a renewable biofuel.

Unlike other biodiesel crops, algae has the potential to be farmed almost anywhere, including on rooftops and land that isn't fertile enough for other crops. Algae also grows without large amounts of fertilizer or energy inputs and, therefore, produces high energy returns. In fact, some studies indicate that algae ponds can produce more than 100 times more fuel per area than traditional biofuel crops.

Ethanol is a liquid biofuel made from corn, sugarcane, or switchgrass. For many decades, gas companies in the U.S. have been adding ethanol made from corn to gasoline to improve the efficiency of automobile engines. Unfortunately, however, transforming crops into ethanol uses large amounts of water, and overall, the process isn't very energy efficient. As a matter of fact, more energy currently goes into producing ethanol from corn than the ethanol produces itself! Not to mention, creating fuel from crops like corn requires all the inputs of fertilizer, pesticides, and fossil fuels needed to grow those crops.

Brazil has created a sustainable fuel economy based on sugarcane ethanol and flex-fuel vehicles. *Flex-fuel vehicles* are designed to run on fuel that contains a combination of ethanol and gasoline. Brazilians use sugarcane, which requires less fossil fuel energy to farm and harvest than corn, so the energy returns on sugarcane ethanol are much greater than they are on corn ethanol. Many people consider Brazil's success with biofuels an example of how to develop sustainable energy economies in other nations.

Unfortunately, the production of the plant matter and oils used to create biofuels relies on agricultural resources and may compete with food production in some regions. To deal with this problem, some scientists have developed ways to convert used cooking oils (such as from fast-food restaurants) into biodiesel fuel for car engines. They've also searched for ways to grow biofuel crops efficiently without impacting food supply.

Methane gas

When bacteria break down (or digest) biological matter in environments that have no oxygen (like wetlands, landfills, and cow stomachs), they produce methane gas. When methane gas is burned, it combines with oxygen in the air, releases energy, and produces CO_2 and water as byproducts, which means it's a relatively clean fuel.

REMEMBER

One of the biggest advantages of using methane as a fuel source is that it removes methane from the atmosphere. Why is that so important? Because methane is a powerful greenhouse gas that leads to global warming; see Chapter 19 for more details.

The biggest challenge to using methane energy is that it's difficult to store and transport in gas form. Another issue is that the technology currently used to convert methane into energy isn't cheap enough to convince people to switch to methane from fossil fuels. However, as global fossil fuel resources decline, technological innovation is likely to produce more cost-effective approaches to methane fuels.

Harnessing Energy from Water

Some of the most plentiful sources of alternative energy are based on the movements of water. Water in motion, as it flows down hillsides or crashes as waves on the beach, for example, is a form of energy called *kinetic energy* (see Chapter 4). Converting this kinetic energy into electric energy is already a common practice in many parts of the world.

In this section, I describe how humans convert the flow of rivers into electricity through hydropower dams and how new technologies help harness the power of ocean waves and tides. The best part about both of these water sources is that they're 100 percent renewable; the supply of water on the land surface is constantly being replenished through the sun-driven process of evaporation in the hydrologic cycle (see Chapter 6).

Damming rivers: Hydropower

All rivers and streams flow downhill across the land surface. Humans can convert this motion of water, which is a form of kinetic energy, into electricity by building dams across rivers. Each dam directs the water flow of the river through a system that generates electricity. Figure 14-1 shows a simple hydropower dam system.

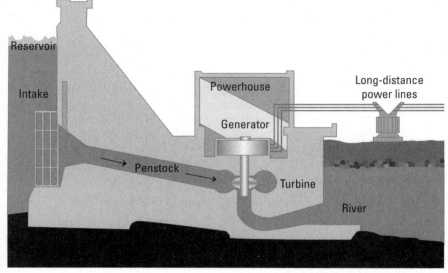

Figure 14-1:
A
hydropower
dam that
generates
electricity.

So how exactly does a dam generate electricity? The dam creates a large lake, or *reservoir* of water. The river water then flows from the reservoir through the dam's *penstock* (control gate or pipe) into a *turbine* or large water wheel. As the water pushes the turbine, the kinetic energy is transferred into a generator that converts it into electric energy. Electric companies can then send the electric energy along wires to provide power to industries or households. Many regions around the world currently depend on hydropower dams to produce electricity.

The Three Gorges Dam in China illustrates both the advantages and the disadvantages of using hydropower energy. The main advantages are that the dam provides China with huge amounts of renewable, clean energy with low monetary costs (other than the initial construction costs). Unfortunately, though, when the Three Gorges Dam was built, it flooded extensive watershed regions, drowning whole villages, whose populations had to relocate above the reservoir waterline. The building of the dam also destroyed millions of acres of habitat and, thus, had negative effects on biodiversity. The verdict is still out on whether meeting the growing energy needs of China's people through a clean and renewable energy source (hydropower) is worth the destruction of habitat and the other negative outcomes.

Because many people criticize large hydropower dams for the environmental damage they cause (such as destroying habitat and changing the nature of the river ecosystem), some scientists have started developing new technologies that apply the same principals to capture river energy as a traditional dam without causing as much damage. For instance, small floating systems such as *microhydropower dams* and *run-of-the-river systems* capture kinetic energy from flowing rivers without causing dramatic ecosystem disruption. These smaller hydropower systems use tiny turbines to capture the natural flow of water instead of creating a lake or causing flooding. The smallest systems sit under the surface of the water and have very little impact on the environment around them. Small-scale systems such as these are examples of what the future of hydropower may look like.

Feeling the pull: Tidal and wave energy

The oceans are another source of hydropower energy. Twice a day as the Earth rotates, the ocean waters move into and away from the coast in what you know as *tides*. To take advantage of this water flow, scientists have come up with a way to convert the kinetic energy from the tides into electric energy, similar to the way they use river hydropower systems (or dams). When water flows toward the shore with the incoming tide, it flows through a *tidal energy system*, turning a turbine that generates electricity. When the tide shifts and flows in the opposite direction, away from shore, the tidal energy system again captures the energy with a turbine and converts it into electricity.

The main advantage of using ocean tides for energy is that tides and waves are a constant, unchanging, clean source of energy. The main disadvantage is that tidal energy systems may impact the environment by changing the flow of water into shoreline and estuarine ecosystems.

Getting Steamed: Geothermal Energy

The Earth's interior produces heat that humans can capture and use for energy. Heat energy from the Earth is called *geothermal energy,* and although it can't fuel automobiles, it can heat buildings and water. Plus, it's a reliable, clean energy source that's fairly cost-effective after the right energy system is in place.

Geothermal heat pump systems are fairly simple, as shown in Figure 14-2. In a geothermal heat pump system, a pump sends water through pipes deep underground to where the Earth's internal temperature is warmer than it is at the surface. The pump then pushes the warmed water back up to the surface through pipes that can heat your house or heat water for your family's household needs. The only drawbacks of geothermal heat pump systems are that they cost a lot to install and they're not available in all regions of the world.

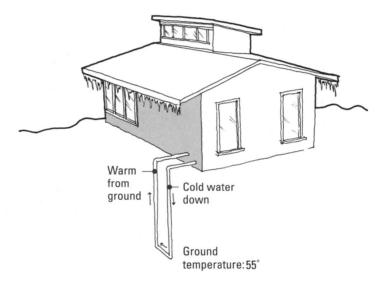

Warm from ground ↑   ↓ Cold water down

Ground temperature: 55°

Figure 14-2: Geothermal heat pump system.

Illustration by Wiley, Composition Services Graphics

Soaking Up the Sun: Solar Energy

Humans can capture solar energy directly from the sun in various ways. The longest-used way is through basic *passive solar energy systems*. For example, ancient people made their houses out of stone or clay, which absorbed the sun's heat during the day and stayed warm after dark, providing heat throughout the night. Builders today use similar methods for passively capturing solar energy. For example, they construct houses with large double- or triple-paned windows that get direct sunlight to capture and magnify the sun's warmth. The effect is similar to but more powerful than what happens to your car on a sunny day: The air inside becomes much warmer than the air outside because the windows let in the sun's energy and trap it, gradually raising the temperature.

Other effective methods of passive solar energy capture include using stone flooring and walls with thick insulation to keep the energy in buildings. With carefully placed windows and other architectural techniques, passive solar energy systems can be an effective way to heat buildings.

Active solar energy systems use the same principles as passive systems except that they use a fluid (such as water) to absorb the heat. A solar collector positioned on the roofs of buildings heats the fluid and then pumps it through a system of pipes to heat the whole building.

Photovoltaic cells, or solar panels, are slightly more involved than passive or active solar energy systems. They convert sunlight to electricity by using thin sheets of silicon. These thin sheets are inexpensive and can be added to roof tiles. People in remote areas such as mountain tops and islands often use photovoltaic cells to generate electricity in their homes and businesses. Figure 14-3 illustrates how solar panels capture sunlight and generate electricity.

The good news about solar energy is that the sun is always available. The bad news is that depending on the system, solar energy may be too expensive for widespread consumer use. Even so, technological advances continue to lower the costs of using solar energy systems for electricity, so that may change in the future.

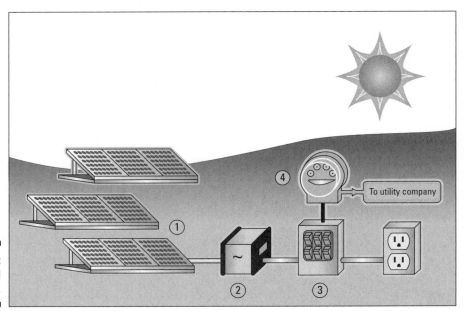

Figure 14-3:
Solar cell
system.

To utility company

Illustration by Wiley, Composition Services Graphics

Chasing Windmills: Wind Energy

Wind energy is another type of alternative energy fueled by the sun. When sunlight enters Earth's atmosphere, it warms the air (see Chapter 7). The warm and cool air in Earth's atmosphere is constantly moving around, creating patterns of wind and weather. The motion of air in wind is similar to the motion of water in rivers and oceans (see the earlier section "Harnessing Energy from Water" for details). And like water, wind is a form of kinetic energy. In certain regions of the world, due to geographical factors and weather patterns, the wind blows strongly and steadily almost all the time. As a result, humans can capture and convert this energy into electricity by using wind turbines.

Wind turbines are like giant pinwheels; as the wind flows against the pinwheel, the arms move, turning a turbine inside that generates electricity. In most areas, humans build multiple wind turbines together, creating *wind farms* or *wind parks*. Recently, people have built turbines along coastlines in what are called *offshore wind farms*. These particular wind farms have two main benefits: They aren't as visible as mainland wind farms, and they capture the energy of the strong winds blowing across the ocean.

The main disadvantages of wind farms are that they produce low-level noise pollution and they can be quite ugly to look at. Plus, some wind farms have resulted in bird, bat, and butterfly deaths when the animals fly into the path of the spinning arms. To deal with this environmental issue, engineers now consider animal migration routes as they plan future wind farm sites and have started designing safer turbine structures.

Wind energy is the fastest-growing alternative energy industry, and it has great potential for supplying the world's electricity needs. In fact, scientists estimate that wind could supply up to half of the electricity needed worldwide.

Energy on Demand: Creating Fuel Cells

A cutting-edge industry that holds a lot of promise for meeting future energy needs involves the creation of hydrogen fuel cells. *Hydrogen fuel cells* combine hydrogen atoms with oxygen atoms to form water and, in the process, generate electricity. Figure 14-4 illustrates a very basic hydrogen fuel cell system.

In the hydrogen fuel cell system, each hydrogen atom is split into a proton and an electron, and the two parts are sent on different paths. The flow of electrons through a wire outside the system creates electricity. The electrons then reenter the system and recombine with the protons as well as oxygen atoms, forming water molecules (H_2O).

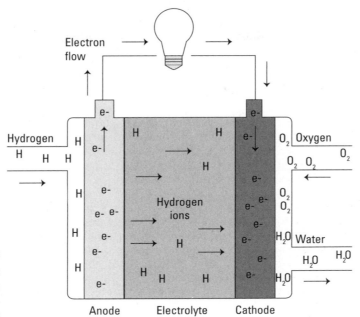

Figure 14-4:
Hydrogen
fuel cell.

Illustration by Wiley, Composition Services Graphics

Hydrogen fuel cells are highly efficient at producing electricity, and their only waste material is water, making them a clean energy source. The tricky part is supplying the pure hydrogen to the cells. Hydrogen gas (H_2) is explosive and most often found in molecules such as water (H_2O), methane (CH_4), and many others. To make hydrogen fuel cells an efficient alternative energy source, scientists need to come up with a way to produce, contain, and transport hydrogen gas that doesn't require large amounts of fossil fuels. Solar and wind power are two options that scientists continue to explore.

Making the Most of the Energy You Already Have: Energy Conservation

The ugly truth is that human beings waste a lot of energy. Although some energy is lost due to the nature of energy transfer and entropy (see Chapter 4), humans waste an almost equal amount (nearly 40 percent) in the processing, transportation, and overall use of energy.

Energy efficiency or *conservation* refers to the process of reducing the amount of energy that's wasted, and some scientists consider it an energy source of its own. Rather than capturing more energy from the sun, water, or fossil fuels, energy efficiency allows you to capture it before it's wasted and to put it to good use.

Humans can save or conserve energy in both large and small ways. In this section, I describe a few of the methods people can use today to help conserve energy, including things you can do at home.

Cogenerating: A two-for-one deal

Industrial plants that produce energy have long been the guiltiest of wasting energy. For example, most fossil fuel-burning plants that produce electricity release large amounts of heat into the environment as an unwanted, wasted byproduct. However, those same plants could capture this heat energy and put it to work through cogeneration technology.

Cogeneration is the production of two types of energy at a single plant. Most commonly, a plant produces both electricity and steam or hot water . Ideally, the plant can use the steam or hot water energy nearby so that it doesn't lose much energy during transport. Cogeneration improves energy efficiency from 30 percent energy capture to almost 90 percent, and it reduces carbon emissions. Instead of using twice the amount of fossil fuels to double the energy output, cogeneration creates a two-for-one energy production deal.

Power plants that are located close to the buildings they supply power to can easily use cogeneration to supply electricity for lighting and hot water or steam for heating. But in the last 50 years, as urbanization and suburban development has expanded across the landscape (see Chapter 10), power suppliers have moved farther and farther away from the buildings they supply as the idea of living or working right next to an industrial power plant has become unappealing to families who enjoy the fresh air and green spaces of a suburban landscape. As a result, using the steam or hot water cogenerated with electricity isn't as cost-effective as it used to be because it requires longer transportation, during which much of the energy is lost.

The good news is that power plants that do use cogeneration today are much cleaner, safer, and less toxic than they used to be. Also, the benefits of reducing carbon emissions and conserving huge amounts of energy outweigh the possible downside of locating them closer to urban areas.

Building smarter power grids

Electricity often gets lost along the journey from where it's produced to where it will be used. Now I'm not talking about getting lost in the sense of needing a map but getting lost in the sense of being wasted and no longer available for use. Even if the production plant generates the electricity efficiently, much of that electricity gets lost as it travels along the electrical grid wires to your house.

You may think you're saving electricity for later when you turn off your lights. But in reality, the production plant supplies the same amount of electricity through the power lines, regardless of how many lights you have turned on. Even if you aren't using the electricity, it will still dissipate due to the inefficiency of power lines and entropy.

The solution to this waste is a smart grid. A *smart grid* is an electricity transportation system that can adjust itself to meet changing energy demands. For example, you probably use more electricity in your home during the hours after you get home from work until you go to bed to run your lights, TV, computer, dishwasher, refrigerator, and so on. This time period is considered a *peak period* of energy use. But when you go to sleep, you turn off or don't use as many of these appliances so they no longer need a steady supply of electricity. This time period is known as the *off-peak period* of energy use.

A smart grid, programmed and controlled by computers, tracks when you need different amounts of electricity and adjusts how much it sends your way according to those needs. Of course, you still have electricity available during off-peak hours, but the grid supplies less of it automatically so that it isn't wasted.

Smart grids can function on a larger scale than just your house. For one, smart grids measure and respond to changes in energy use at the local (neighborhood) and regional scales, such as responding to when the sun sets in Chicago a few hours before it sets in Seattle.

The future of smart grid technology and energy efficiency may include smart meters and smart appliances. *Smart meters* record your energy use in smaller intervals (such as every hour) and share that information with the utility company so it can more efficiently meet your energy needs. Similarly, *smart appliances* are set to turn themselves on and off based on the off-peak (and therefore lower-cost) hours of the day. For example, imagine that you loaded your dishwasher in the morning and left for work knowing that it would turn itself on midday during off-peak hours when electricity costs are the cheapest. Not only would you save energy, but you'd also save money. And this future is just around the corner.

Getting around town more efficiently

One of the biggest uses of energy today is in transportation. The fossil fuel-driven vehicle is a hallmark of modern society, particularly in the U.S., where people have a strong attachment to their cars. Environmental scientists and others have made huge efforts to improve fuel efficiency and reduce dependence on automobiles in the name of climate change, but energy conservation also plays a part in improving transportation efficiency.

Here are some of the ways transportation industries are conserving energy:

✔ Increasing fuel efficiency (getting more miles of travel per gallon of fuel)

✔ Producing *hybrid* vehicles that use both gasoline and electricity for fuel

✔ Developing electric vehicles that need no gasoline at all

The transportation industry is still working toward developing lower-cost hybrid options for the average consumer. Hybrid and electric technologies have already impacted energy conservation in public transportation. For example, many cities use hybrid or electric buses and trains to reduce the need for automobiles. The result is greater transportation energy efficiency and cleaner air.

Capturing energy at home

Although energy conservation by industries and urban mass transit seems to have the biggest impact on overall energy use, don't forget that every bit of energy saved counts. After all, any energy you save helps extend the available energy resources on Earth, lower energy costs, and reduce waste. Here are some ways you can begin conserving energy in your own home today:

✔ Install compact fluorescent light bulbs (CFLs) or LED bulbs around your house.

✔ Turn off electronic devices when you're not using them. Using a power strip that has an on/off switch is a great way to simplify turning off all your gadgets at once.

✔ Install proper insulation in the attic and walls of your home to reduce heat loss during the winter and retain cool air in the summer.

✔ Install a programmable thermostat to automatically lower the setting when you leave for work and raise it before you get home. Also, set your thermostat a little lower in the winter and wear a sweater.

✔ Run only full loads in your dishwasher and clothes washer and air-dry your clothing and dishes when possible.

✔ Replace your furnace filters regularly and maintain clean air ducts and vents.

✔ Check that all your doors and windows are sealed properly. Use caulking and weather stripping to improve the seals.

Part IV
Giving a Hoot: Pollution and Environmental Quality

The 5th Wave   By Rich Tennant

"Now you've really done it, young man. You've poked a hole in the ozone layer!"

In this part . . .

Do you know where your garbage ends up? In this part, I describe the result of many decades of human pollution of the environment. Humans damage water, air, and land resources when toxins leak from their trash or are released into the environment during product manufacturing. This part is eye-opening if you've never considered where all that waste goes after the garbage truck collects it from your curb.

Environmental scientists focus much of their energy on cleaning up this pollution and developing new ways to make products that don't result in so much pollution. Here, you find out more about the current methods of improving environmental quality and reducing pollution of water, air, and land. And you discover how the Earth's climate is changing, how human impacts are affecting the climate, and what must be done in the face of a warmer future.

Chapter 15

Breathing Room: Addressing Manmade Air Pollution

*F*resh, clean air is something many people take for granted. Until the air they're breathing becomes polluted — either darkened or irritating to inhale — folks just don't realize how much they depend on clean, breathable air. The air around you is susceptible to natural and human sources of air pollution that damage air quality and reduce the availability of this important resource.

In this chapter, I describe the common air pollutants and their effects. I focus on manmade air pollution, explaining how pollutants in the air are detrimental to human health and damage ecosystems. I also discuss how manmade air pollutants harm the Earth's ozone layer, and I lay out the effects of indoor air pollution.

Sorting Out Common Pollutants

Scientists sort air pollution into three main categories. *Primary pollutants,* such as the exhaust from an industrial smokestack or your car, enter the air in harmful form. In contrast, *secondary pollutants* enter the atmosphere in a harmless form and then transform (usually by exposure to sunlight) into harmful chemicals. (Sometimes a primary pollutant can also become a secondary pollutant after exposure to sunlight.) The ozone in smog is a secondary pollutant that forms when gasoline vapors react with sunlight.

A third category of pollutants doesn't come from a smokestack or exhaust pipe. *Fugitive emissions* occur when pollutants leak or are inadvertently added to the atmosphere. Many fugitive emissions are simply leaks, but this category also includes dust, which is churned up by human activities such as agriculture, construction, rock crushing, and mining. Although dust isn't necessarily poisonous, breathing it can cause you respiratory irritation. Dust affects environmental air quality by darkening the sky and blocking the sun, which impacts photosynthesis (see Chapter 4).

Regardless of whether the pollutants enter the air as primary, secondary, or fugitive emissions, scientists recognize that certain chemical compounds, called *conventional pollutants,* make up the majority of air pollutants today. The following list describes these conventional pollutants:

- **Sulfur compounds:** Burning coal and oil unleashes sulfur compounds such as sulfur dioxide (SO_2) into the atmosphere; volcanoes and forest fires also release small amounts of sulfur dioxide. Sulfur compounds bond easily with water molecules in the atmosphere, forming acid rain (which I describe later in the section "Melting monuments: Acid rain"). Breathing air polluted with sulfur can cause respiratory irritation and illness. Sulfur dioxide in the air also damages plant leaves and disrupts photosynthesis.

- **Nitrogen compounds:** Nitrogen compounds are molecules that include nitrogen — most commonly, nitrogen oxides with one or two oxygen molecules (NO or NO_2). Burning coal and wood releases nitrogen into the atmosphere; the nitrogen then bonds with the oxygen already there to form these compounds. One of the greatest dangers of producing nitrogen compounds in the atmosphere is that they bond with water, forming acid rain.

- **Carbon compounds:** Burning fossil fuels or any other organic materials (including wood, charcoal, and manure) emits carbon dioxide (CO_2) and carbon monoxide (CO) into the air. Carbon monoxide is colorless and odorless, but when humans or animals inhale it, it disrupts their normal respiration processes and can lead to death. Carbon dioxide is the gas that plants use during photosynthesis. The increased use of fossil fuels in the last century has caused CO_2 to become a major pollutant contributing to climate change (more details on climate change in Chapter 19).

✔ **Particulates:** Some air pollutants are larger than single molecules. *Particulates* are small particles of solid or liquid material suspended in the air (sometimes they're also called *aerosols*). Particulates range in size from 0.01 micrometers up to 100 micrometers (for reference, a human hair is about 50 micrometers wide on average). Your nose and throat filter larger particulates out of the air when you breathe, but smaller ones can get into your lungs and cause irritation. And the smallest particulates are often the most toxic substances and can create serious health problems when inhaled. Particulates also cause an environmental problem because they block sunlight from reaching plants on the ground, which means the plants can't undergo photosynthesis.

✔ **Metals:** Some metals enter the air from industrial processes and the burning of fossils fuels. Lead used to be a gasoline additive that improved engine performance; however, it also entered the atmosphere with car exhaust. Mercury is a side product of burning coal and oil; in fact, coal-burning electricity plants are the largest source of mercury pollutants in the air. Waste burning at industrial incinerators also releases mercury. Metals are particularly dangerous because of how they *biomagnify,* becoming more concentrated at the top of the food chain (see Chapter 17 for details).

✔ **Volatile organic compounds:** *Volatile organic compounds,* or VOCs, are organic molecules that occur in gas form at normal atmospheric temperatures. Like any organic compound, VOCs come from both natural and manmade sources. The most common VOCs are *hydrocarbons* — molecules of bonded hydrogen and carbon atoms — and many, such as paints, perfumes, gasoline, and lighter fluid, give off strong odors. VOCs are common indoor air pollutants, originating from household products such as cleaning solutions and cosmetics.

✔ **Photochemical oxidants:** *Photochemical oxidants* are benign compounds that transform into harmful substances after they encounter sunlight (*photo* means "light"). Most secondary pollutants are photochemical oxidants; the most common is ozone (O_3). Although ozone occurs naturally in upper levels of the atmosphere, it occurs as a secondary pollutant — a photochemical oxidant — in the lower layer (the troposphere). Tropospheric ozone pollution is created when sulfur and nitrogen compounds are exposed to sunlight and transformed. Photochemical oxidants play an important role in forming smog, which I explain in the later section "Trapping pollutants: Temperature inversions and smog."

Table 15-1 summarizes the major air pollutants, their sources, and their effects on humans and ecosystems.

Table 15-1	Conventional Air Pollutants	
Pollutant	*Source*	*Impact*
Sulfur compounds	Burning coal, oil, and gasoline	Respiratory illness, plant damage, acid rain
Nitrogen compounds	Burning fossil fuels, biomass, and wood	Respiratory illness, smog formation, acid rain
Carbon monoxide	Incomplete burning of anything	Headache and possible death (at high levels)
Particulates	Burning of materials; agriculture; erosion; construction	Respiratory illness, haze, smog
Lead	Gasoline, oil	Toxic effects on nervous systems of living things
Mercury	Burning coal and oil; mining	Bioaccumulation in the food chain, toxic effects on nervous system
VOCs	Evaporation of fuel, solvents, paint, and gasoline	Creation of tropospheric ozone
Carbon dioxide	Burning fossil fuels	Increase in climate warming

Observing the Effects of Air Pollution

In a nutshell, *air pollution* is the result of adding compounds or particles to the air that are harmful to human health or the environment. The most obvious danger of air pollution is that humans and other animals inhale pollutants and can become ill. In particular, air pollution leads to lung and respiratory illnesses — including asthma, bronchitis, and emphysema — and various cancers. Adding pollutants to the atmosphere damages the environment as well as human health. In this section, I describe a few of the most common results of air pollution, including ecosystem damage, smog, and acid rain.

Losing vegetation

Chemical pollutants in the air directly affect plants, including agricultural crops, that absorb CO_2 gas through their leaves. Absorbing high levels of sulfur compounds from the air can destroy plant cells.

Some studies suggest that even when plants aren't visibly damaged by air pollutants, the overall productivity of plants is lower as a result of pollutants in the atmosphere. Understanding how air pollution decreases plant productivity is particularly important for farmers, who strive to produce maximum yield from each harvest.

Tinting the sky a hazy shade of brown

The sky is *hazy* anytime dust, pollutants, smoke, or other particles change your ability to see clearly. Clouds of haze are most common after forest fires or near industrial sites that emit pollutants into the air. But in recent decades, places far from cities and industry have become hazy — evidence that air pollution circulates far from its source and affects environments that are thought to be a safe distance from pollution sources.

Grand Canyon National Park in the U.S. is one place where haze has increased in the last 50 years. Although haze is troublesome because it indicates the presence of dangerous air pollutants, it also diminishes the beauty and aesthetic value of a wilderness landscape like the one at the canyon. This type of environmental effect is difficult to quantify, but it's an important consideration in calculating the cost of air pollution.

Trapping pollutants: Temperature inversion and smog

You may be familiar with the term *smog* to describe the hazy brown color in the air around large cities or after a forest fire. The term originally described the specific combination of smoke and fog that discolored the air as a result of coal burning during the Industrial Revolution, but *smog* no longer means simply smoke and fog. Nowadays, the term *smog* refers to the complex combination of primary and secondary pollutants that turn the air a brown or yellow color. (The earlier section "Sorting Out Common Pollutants" gives you the skinny on the different kinds of pollutants.)

Although smog isn't isolated to urban areas, it's more common around cities. One factor that intensifies smog in urban areas is the occurrence of a *thermal inversion* or *temperature inversion* in the atmosphere. Generally speaking, the temperature of the air becomes gradually cooler as you move upward in the atmosphere; warm air near the surface moves upward, gradually cooling. In the case of a temperature inversion, however, atmospheric circulation and geographic factors trap a layer of warm air between two layers of cooler air, inverting or flipping the usual pattern. Figure 15-1 illustrates what this effect looks like.

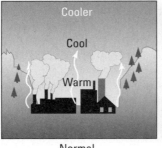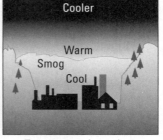

Normal   Temperature inversion

Figure 15-1:
A
temperature
inversion.

Illustration by Wiley, Composition Services Graphics

Temperature inversions are most common in valleys where cool mountain air sweeps down into the valley at night, below the warm, polluted air surrounding the city. Los Angeles commonly experiences temperature inversion that traps a smog layer with cool air above and cool air below. This inversion keeps the smog (and therefore the pollutants) close to the ground instead of allowing it to disperse into the atmosphere. Cities that experience such trapped smog may issue local air quality warnings so that people with asthma or other respiratory troubles know to stay indoors.

Melting monuments: Acid rain

Air pollution creates conditions in the atmosphere that change the pH of rainwater (and snow and other precipitation), making it more acidic. This *acid rain* is an environmental hazard because acidic rainwater damages whatever it falls on. (In Chapter 3, I explain how *pH*, or the acidity of a liquid, can vary.)

People first noticed the damage caused by acid rain more than 100 years ago. Folks in the mid-19th century observed that rainfall in heavily polluted cities (such as London during the Industrial Revolution) was dissolving marble and limestone statues. Not until the 1960s did scientists begin to carefully study the sequence of chemical reactions that results in acid rain.

By far, the worst ecosystem damage acid rain causes occurs in aquatic and wetland ecosystems. Acid rain creates acidic conditions in lakes, ponds, rivers, and wetlands, and the organisms in these ecosystems just aren't adapted to survive in that kind of environment. *Acidification,* or increasing the acidity (lowering the pH) of an aquatic ecosystem, can kill aquatic organisms such as fish and amphibians as well as interfere with their life cycles.

Normal atmospheric water has a pH of approximately 5.6 because of the natural formation of carbonic acid from atmospheric carbon dioxide gas. Pollutants such as nitrogen oxides and sulfur compounds in the atmosphere create atmospheric water particles that are more acidic than normal with pH values less than 5.

Changing the pH of aquatic ecosystems may also have the side effect of allowing other contaminants (such as metals or toxins) to dissolve into the water and move more freely around the environment.

Holy Ozone! Remembering the Hole in the Ozone Layer

Air pollution has the ability to affect the Earth's entire ecosystem, as scientists realized when they found direct links between ozone layer destruction and manmade pollutants called CFCs. In this section, I describe the ozone layer, explaining why it's important for maintaining life on Earth and how air pollution continues to damage this protective layer.

Before you can get a true handle on the ozone layer, though, you need to be familiar with the atmosphere as a whole. Earth is surrounded by layers of gas with different temperatures, pressures, and compositions. Figure 15-2 illustrates each of the layers in Earth's atmosphere.

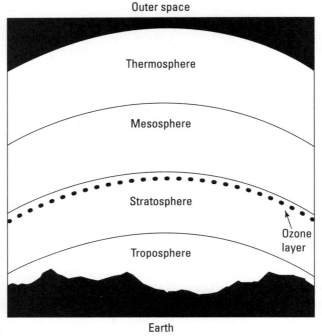

Figure 15-2:
The layers
of Earth's
atmosphere.

Illustration by Wiley, Composition Services Graphics

Starting at ground level, the layers proceed up as follows:

1. Troposphere

2. Stratosphere

3. Mesosphere

4. Thermosphere

So where's the ozone layer in this list? Actually, the ozone layer isn't a specific level of the atmosphere; rather, it's a thick band of ozone molecules at the top of the stratosphere level.

Creating ozone naturally

The Earth's *ozone layer* is composed of molecules of *ozone,* or O_3. These molecules provide a protective layer around the Earth, absorbing a large portion of the sun's radiation, particularly some of the harmful ultraviolet (UV) rays that cause sunburn.

Ozone is naturally created when UV radiation hits oxygen gas molecules (O_2) and breaks some of them apart into single oxygen atoms. These single atoms of oxygen then bond with other O_2 gas molecules, forming O_3. In a similar reaction, O_3 molecules can be broken down into atoms of O and molecules of O_2. In this way, ozone, oxygen gas, and free oxygen atoms are constantly being combined, broken apart, and recombined in the stratosphere. These reactions are driven by the energy from UV radiation, which is absorbed in the process.

Figure 15-3 and the following formulas illustrate this process:

$$O_2 + O + \text{UV energy} \rightarrow O_3$$
$$O_3 + \text{UV energy} \rightarrow O_2 + O$$

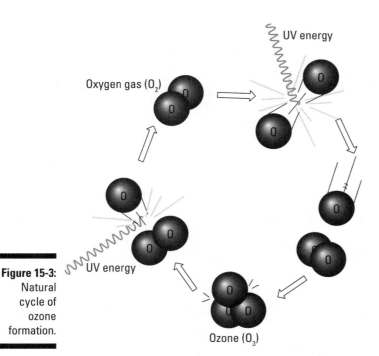

Figure 15-3: Natural cycle of ozone formation.

Illustration by Wiley, Composition Services Graphics

Molecules of ozone found closer to the Earth's surface (in the troposphere) are considered a pollutant. In the troposphere, ozone gas can damage lung tissue and plants. But in the stratosphere, ozone gas provides valuable protection by absorbing harmful UV rays. Because of the intensity of the sun's UV radiation, no living things could survive on Earth's surface without this stratospheric ozone layer.

Depleting the ozone layer

In the 1980s, scientists noticed that the ozone layer over the South Pole was becoming much thinner than usual. After some study, atmospheric scientists recognized that this *ozone depletion* was a result of air pollutants from human activities — specifically, the release of nontoxic, aerosol gas molecules called chlorofluorocarbons.

Chlorofluorocarbons, or CFCs, are chemically inactive (inert) gas molecules commonly used in fire extinguishers, many different types of aerosol cans (such as hairspray), and refrigerator coolants beginning in the 1920s. (You may have heard of Freon, which is the trade name for a specific coolant CFC sold by DuPont.) Using such products releases CFCs into the atmosphere, an occurrence no one worried too much about for many decades; after all, the molecules are chemically inert and nontoxic, so no problem, right? Not so fast.

After the stable and inert CFCs are in the atmosphere, they don't break down or react with anything around them until they reach the stratosphere, where the UV radiation is very intense. The high levels of energy in the UV radiation are strong enough to break chlorine atoms away from the CFC molecule. These free chlorine atoms then break apart ozone molecules in the following reaction:

$$O_3 + Cl \rightarrow ClO + O_2$$

Although the third oxygen atom in the O_3 molecule is bonded to the chlorine atom, it's more attracted to other, single oxygen atoms, so after the ClO compound is created, it quickly changes again, like this:

$$ClO + O \rightarrow Cl + O_2$$

In these reactions, the chlorine from the CFC molecule is acting as a *catalyst* — it's helping to break apart O_3 molecules and forming O_2 molecules. The result is that, unlike the natural cycle of ozone formation and recycling, the ozone is being broken down into O_2 and not recycled back into O_3 because no free oxygen atoms are left for O_2 to combine with. Figure 15-4 shows this process.

The presence of chlorine atoms in the atmosphere changes the natural ozone reaction (the one that absorbs UV energy) reaction from

$$O_2 + O \leftrightarrow O_3$$

to

$$O_3 + O \rightarrow 2O_2$$

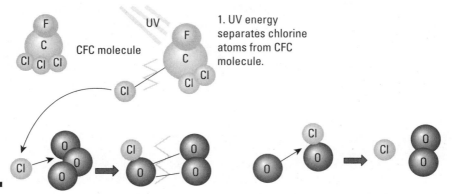

1. UV energy separates chlorine atoms from CFC molecule.

CFC molecule

Figure 15-4:
Ozone
depletion
from CFCs.

2. Free chlorine pulls oxygen from ozone molecule, creating ClO and O_2.

3. Free oxygen bonds with oxygen in ClO, leaving a free chlorine atom and a molecule of O_2.

4. Free chlorine atom is available to break apart more ozone molecules.

Illustration by Wiley, Composition Services Graphics

The second reaction is driven by the presence of chlorine atoms and doesn't absorb UV energy. The result is a stratosphere lacking in ozone molecules; because of this change, the protective layer of ozone has gradually become thinner and less protective all over the world.

Scientists studying ozone depletion have also noticed that the thinning ozone layer at the South Pole is at its thinnest during the Antarctic spring (September and October), creating an ozone hole when the sun returns after the dark winter. (Ozone depletion also occurs over the Arctic in the Northern Hemisphere spring, but it isn't as extreme as the Antarctic version.) What scientists have realized is that ice crystals containing nitrogen oxide molecules from air pollutants form during the cold winter, providing a surface for chlorine molecules to attach themselves to. In the spring, when the sun returns and the atmosphere warms, the ice melts and releases the chlorine into the stratosphere, where it interferes with ozone formation, resulting in the ozone hole.

Halting ozone depletion: The Montreal Protocol

When scientists and the rest of the world realized that human actions and industrial air pollution were responsible for ozone depletion, the international community decided to solve the problem together.

In 1987, world leaders met at a conference in Montreal to discuss solutions to ozone depletion. The result of this international meeting of the minds was the *Montreal Protocol,* which outlined specific ways for international communities, businesses, and industries to reduce CFC output. Their original goal was to stop producing CFCs by 2000. They later moved this deadline to 1996 as more scientific evidence illustrated the rate and degree of damage being done.

The success of the Montreal Protocol in reducing CFCs was an inspiring display of international cooperation and commitment to global environmental preservation. Different chemicals have been developed to fill the role of CFCs without releasing the ozone-damaging chlorine, and the international community continues to help poorer nations transition to new technologies and reduce CFC output.

The good news is that CFC production has been reduced by 95 percent and that stratospheric ozone levels have rebounded and are expected to reach their previous (pre-CFC) levels by the mid-21st century. The bad news is that the ozone is still being depleted as a side effect of modern climate warming, which I describe in Chapter 19.

The Air in There: Watching Out for Indoor Air Pollution

Although air pollution damages ecosystem health and can endanger human health, the greatest danger to humans from air contaminants actually occurs indoors. You may consider the inside of your home the safest place for you and your family to avoid air pollution. Unfortunately, air pollutants fill the air inside most homes and affect residents' health. Some of the most common air contaminants in your home (not including cigarette smoke) are

- Chemicals released from building materials, flooring, and furniture (such as formaldehyde and asbestos)
- Mold spores, bacteria, pet dander, dust, and other irritating particulates
- VOCs from cleaning products, paints, and other solvents (such as paint thinner)
- Sulfur, nitrogen, and carbon compounds created during wood burning in stoves and fireplaces and fossil fuel burning in cars

The air pollutants in your home are no different from the pollutants found outside your home, except that indoor pollutants concentrate to dangerous levels more quickly. In countries where families cook over small fires or open flames indoors, the amount of carbon monoxide and particulates in the indoor air is much higher. In more developed countries, the toxic gases from building and furniture materials are the biggest source of indoor pollution.

The health effects of indoor air pollution have led scientists and doctors to recognize *sick building syndrome.* People experiencing sick building syndrome display symptoms that aren't directly connected to an illness or disease. The symptoms typically occur because the indoor air contaminants build up in areas with poor ventilation or air circulation; they go away when the person leaves the building or room. Symptoms of sick building syndrome include the following:

- Headache
- Respiratory irritation
- Dizziness
- Skin irritation
- Sensitivity to odors
- Nausea
- Fatigue

Improving the air circulation and ventilation is the first step to making indoor air safer. Other solutions include using air filters to clean the air and performing intensive cleaning to remove pollutant sources (such as hidden mold growth or carpet that collects dust, mold, and pollen).

Clearing the Air

The first step in reducing air pollution is to minimize production of primary pollutants and compounds that become secondary pollutants. (You can read about these designations in the earlier section "Sorting Out Common Pollutants.") Energy conservation in industry and transportation goes a long way in reducing the number of pollutants released into the atmosphere. To complement conservation methods, new technologies have also been developed that offer promising solutions to pollution-related environmental problems. These advancements include filter systems and air scrubbers that remove particles from dirty air.

Laws and regulations help to enforce conservation approaches, which recognize that the easiest place to clean polluted air is before it enters the atmosphere, and to hold industry responsible for the pollutants it produces. I describe some of the most important air pollution legislation (such as the Clean Air Act) in Chapter 21.

Chapter 16

Drip Drop Splash: Water Pollution

*W*ater covers more than 70 percent of the Earth's surface. In fact, this natural resource is so vast that for many decades, as the Industrial Revolution blossomed, humans simply dumped waste materials into the nearest body of water under the assumption that the water would carry them far away or dilute them to meaningless levels.

Over the last 60 years, however, people have realized that water, like other matter, cycles through the environment (see Chapter 6 for details on the water, or hydrologic, cycle). This means that anything added to water in one location eventually affects the water resources of living things (plants, people, and other animals) in other locations.

The good news is that because of action and regulation of industry, many of the world's rivers are much cleaner today than they used to be. In this chapter, I explain how human waste, industry, and agriculture pollute water, and I describe the most common types of pollution and the many negative effects they have on human health and ecosystems. I also outline some of the traditional ways to combat water pollution and some new approaches that offer sustainable solutions to the environmental challenge of water pollution.

Polluting Water in Many Ways

Water pollution is the result of adding anything to water that diminishes its quality. While you can easily see that throwing empty soda cans into a lake is a form of pollution, most forms of water pollution aren't that visible. Here are the different forms water pollution can take:

✔ **Solid waste:** As you may have guessed, solid waste consists of physical objects, or trash, that humans dispose of in water. The oceans experience a high degree of solid waste pollution, which I describe in more detail in Chapter 18.

✔ **Noise:** Manmade noises, such as from boats, ships, and submarines, are another type of water pollution because these noises degrade the water quality for organisms that live in the water. Scientists are particularly concerned about animals such as whales, whose communication with each other can be disrupted by underwater noise pollution.

✔ **Sediment:** To some degree, natural processes wash some *sediment* (particles of sand, silt, or clay) into water. However, human activities such as agriculture and animal herding greatly increase the amounts of sediment that wash into nearby water through the process of erosion. This addition of sediment blocks sunlight from reaching plants and animals below the water's surface and may even clog the *gills* (breathing organs) of some fish.

✔ **Heat:** The addition of heat to water is called *thermal pollution.* Thermal pollution occurs when humans use water for industrial processes that heat it up and then return it to its source. The organisms in the water ecosystem are adapted to the original water temperature and can experience *thermal shock* because their bodies can't adjust to the increased temperature and they suffocate.

✔ **Toxins and pathogens:** Substances that endanger the health of living organisms include poisonous chemicals, or *toxins,* and disease-causing organisms called *pathogens.* Most toxins that pollute water originate from industrial wastewater or air pollution (mercury and arsenic, for example). Dangerous pathogens enter water supplies via untreated sewage and animal fecal matter. These types of water pollution are the most commonly studied and regulated; I cover them in detail in the section "Endangering Human Health" later in this chapter, as well as in Chapter 17.

Getting to the Point (And Nonpoint) Source

Environmental scientists classify the source of water pollution in one of two ways:

✔ **Point source water pollution:** This type of pollution originates at a specific and identifiable location, such as a drainage pipe. Point source pollution comes from drain pipes at factories, power plants, and other industrial locations, as well as from sewage pipes and drainage ditches.

✔ **Nonpoint source water pollution:** This type of pollution comes from a large area of polluted surface runoff with no specific origination point. The most common source of nonpoint source pollution is runoff from agricultural fields and urban areas. For example, anytime it rains, the rain water washes pollutants from streets and parking lots into nearby streams and rivers, resulting in nonpoint source pollution.

Generally speaking, point source pollution is easier to regulate or control because you know the source of the pollution. You can locate the outlet of a pipe that's causing pollution and divert it to a treatment system to remove pollutants. Nonpoint source pollution is much harder to control because it doesn't come from a particular place and often occurs periodically (such as when it rains) rather than more steadily as is the case with point source pollution.

Endangering Human Health

One of the greatest dangers to human health is water pollution. After all, people can't survive without drinking water, and if their freshwater resources are polluted, they can become ill from drinking them. Different types of pollutants affect human health in different ways. In this section, I explain how pollutants damage human health.

Bacteria, viruses, and parasites, oh my!

Organisms that cause disease are called *pathogens*. Pathogens include bacteria, viruses, and parasitic organisms that infect humans and cause illness. Some pathogens occur naturally, and others pollute water when human or animal waste washes into the water. Some of the most common illnesses caused by pathogens in water include

✔ Typhoid

✔ Cholera

✔ Dysentery

✔ Polio

✔ Hepatitis

These illnesses are particularly dangerous for young children; in fact, they account for almost 60 percent of early childhood (younger than age 5) deaths worldwide. Although sewage treatment plants have reduced the occurrence of water-related illnesses in some nations, less developed nations still struggle to find safe, fresh water that's free of such pollutants. In some regions of the world (parts of India, China, and Africa, for example), water-related illnesses are still a leading cause of death.

In some cases, organisms rather than the water itself carry pathogens. For example, insects whose eggs or larvae live in water carry malaria. Still other diseases are caused by microscopic organisms in the water that infect humans as intestinal parasites — leading to diarrhea, fever, and sometimes death.

Scientists seek to stop waterborne pathogens before they can infect humans and cause illness. Because the pathogens themselves are so difficult to detect, scientists look for an indicator, such as traces of fecal coliform bacteria, instead. *Fecal coliform bacteria,* such as *Escherichia coli,* or *E. coli,* are a common and usually harmless type of bacteria found in human intestines. But if scientists detect *E. coli* in a water source, it indicates that some human or animal waste is present in the water and that the likelihood of other dangerous bacteria being present is much greater.

Pesticides, drugs, and metals

Some water pollutants don't directly cause illness, but they do damage human health (and the health of other organisms) over the long term. These pollutants, called *chemical pollutants,* include manmade organic compounds that humans use to make pesticides, prescription drugs, plastics, and other products.

Chemical pollutants enter water sources as runoff from agricultural fields (pesticides) or as drain water (from kitchens and bathrooms) from human homes and businesses. These pollutants also seep into groundwater reservoirs from landfills and underground sewage containers. After these chemicals enter groundwater sources, they contaminate freshwater drinking supplies and are difficult to clean up.

Chemical pollutants usually occur in very low amounts in water supplies, but even at such low levels, they're still dangerous to human health. Some of these compounds are, at the molecular level, very similar to human hormones and are called *environmental estrogens* or *endocrine disruptors* (more details on these in Chapter 17).

Metals, such as mercury, iron, and nickel, pollute water as well. Some of these metals wash into water during mining operations, whereas others, such as mercury, settle into water via the air after being emitted from industrial smokestacks. The small amounts of these metals that scientists measure in water supplies appear almost harmless. However, after an animal consumes these toxins, they get concentrated and biomagnified up the food chain.

(*Biomagnification* basically means that the negative effects of the toxin are magnified in organisms at the top of the food chain; see Chapter 17 for more details.)

Disrupting Ecosystems

Not only is water pollution dangerous for human health, but it also damages ecosystems. Of course, toxins are dangerous to living organisms, but human water pollution affects ecosystems in other less obvious ways, too. I describe two of these — the creation of dead zones and nutrient pollution — in this section. But I start by describing how scientists measure dissolved oxygen to determine when an ecosystem is being disrupted by pollutants.

Measuring dissolved oxygen

To understand the changes that occur when pollution enters an aquatic ecosystem, scientists measure the amount of oxygen that's dissolved in the water. The *dissolved oxygen content,* or *DOC,* of water helps determine what type of organisms a particular ecosystem can support. If the water doesn't contain enough dissolved oxygen, then the ecosystem can support only organisms that are adapted to low levels of oxygen, which makes for fewer organisms in the ecosystem overall.

Scientists measure dissolved oxygen as the number of oxygen molecules dissolved in water, using the unit *parts per million* or *ppm*. On average, water can hold only about 12 dissolved molecules of oxygen per million molecules of water, or 12 ppm of dissolved oxygen. For most fish to survive, water needs at least 5 ppm of oxygen. Anything less than 2 ppm can support only certain bacteria and other decomposers.

Water gains oxygen in two ways:

- ✔ **By mixing with the air above it:** This means that turbulent whitewater rapids have a much higher DOC than slow-moving or still water.

- ✔ **By housing photosynthesizing organisms:** Plants and photosynthetic algae in the water produce oxygen.

Temperature also largely influences how much oxygen can be dissolved in water; cold water dissolves more oxygen than warm water.

Water loses oxygen when animals in the water use oxygen during respiration and when decomposers use it to break down organic matter. The amount of oxygen needed by these organisms is called the *biochemical oxygen demand,* or *BOD.*

Similar to the BOD, the *chemical oxygen demand,* or *COD,* measures how much oxygen is needed to oxidize (or break down) all the organic and inorganic matter present in the stream (see Chapter 3 for details on oxidation).

Creating zones in aquatic ecosystems

Environmental scientists have observed that when pollutants, particularly sewage, which is full of organic waste matter, enter a flowing body of water (such as a river or stream), a sequence of changes occurs in the aquatic ecosystem downstream from the pollution source. These changes create a series of zones in the water, which you can see illustrated in Figure 16-1.

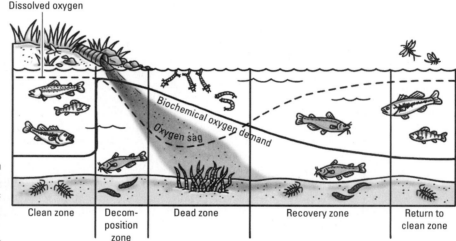

Figure 16-1:
Zones of a polluted waterway.

Illustration by Lisa Reed

Each zone displays certain characteristics of water chemistry and ecosystem health:

- **Clean zone:** Upstream — above where the pollutants enter the water — is the *clean zone.* This zone has high levels of dissolved oxygen in the water and supports a diverse community of organisms at every level of the food chain. Biochemical oxygen demand is well balanced by the presence of dissolved oxygen.

- **Decomposition zone:** At the point where sewage enters the stream, biochemical oxygen demand spikes upward as decomposers consume the organic waste material. The increased consumption of dissolved oxygen removes much of the available oxygen from the water. As a result, the *decomposition zone* supports mostly organisms that survive by eating decomposing matter and organic waste material and that don't need high levels of oxygen.

✔ **Dead zone:** As a result of the dramatic increase in oxygen use in the decomposition zone, the next zone, called the *dead zone,* can no longer support organisms that need oxygen. This zone is where the most extreme *oxygen sag,* or depleted levels of dissolved oxygen in the water, occurs. (This zone is sometimes called the *hypoxic zone* because of its low levels of dissolved oxygen.) In this zone, only organisms, such as sludge worms and certain bacteria, that can survive on very little oxygen exist.

✔ **Recovery zone:** Further downstream, where decomposers finish consuming the waste material and dissolved oxygen is slowly added as the water mixes with the atmosphere above, the ecosystem begins to recover. The *recovery zone* is often the largest zone and may stretch for miles past the point source. In this zone, the biochemical oxygen demand decreases, while the dissolved oxygen levels increase back toward normal levels.

✔ **Return to clean zone:** After the ecosystem has recovered from the waste input, it begins to support the wide variety of organisms that existed upstream from the pollution. This *return to clean zone* has normal levels of dissolved oxygen and biochemical oxygen demand and supports a thriving and diverse ecosystem.

Growing to extremes: Nutrient pollution

A common way that humans pollute water is through the addition of fertilizers and sewage to water as nonpoint source pollution. These added materials are full of nitrogen and phosphorus, two nutrients that encourage the growth of aquatic producers, such as algae. The excessive growth of algae dramatically disrupts the entire ecosystem. Because this type of pollution results from excess nutrients, it's called *nutrient pollution*.

Bodies of water naturally exist with different nutrient levels:

✔ Waters that have low levels of nutrients and, therefore, low numbers of producers and low biological productivity are called *oligotrophic*. Oligotrophic waters are clear without many microscopic algae plants.

✔ Waters that have high levels of nutrients and, therefore, high biological productivity with large algae populations are called *eutrophic*.

The transition from oligotrophic to eutrophic — called *eutrophication* — occurs naturally over time as a type of community succession in lakes and ponds (see Chapter 8). Additional sediment and nutrients enter the ecosystem via streams, rainwater runoff, and other sources. Eventually, the biological activity and plant growth in the lake or pond increase to a degree that it may no longer be a lake or pond, becoming instead a wetland.

When humans pollute water with nutrients, this normally long, slow process of eutrophication occurs very rapidly and is called *cultural eutrophication*. The rate of cultural eutrophication is much faster than natural rates of eutrophication; as a result, the organisms in the ecosystem don't have time to adapt to the changes and die. Here's what the cultural eutrophication process looks like:

1. Pollution enters the water, adding sediment, nitrogen, and phosphorus (the nutrients).

2. The added nutrients fuel algae growth; algae begin to reproduce in huge numbers, creating what's often called an *algae bloom*.

3. The algae population continues to grow until it runs out of either nutrients or space — covering the surface of the water completely.

4. Extensive algae across the surface of the water blocks sunlight from reaching organisms below the surface of the water, such as aquatic plants.

5. Underwater plants begin to die from lack of sunlight, while the large algae population, having reached its limit, also begins to die, sinking into the water where bacteria decompose it.

6. Decomposition of dead plants and algae by bacteria consumes the dissolved oxygen in the water, creating oxygen sag conditions.

7. Fish and other oxygen-breathing organisms suffocate from lack of oxygen, and the entire aquatic ecosystem collapses, creating one big dead zone (see the preceding section).

Cultural eutrophication doesn't occur only in lakes. Coastal waters also experience the effects of nutrient pollution and cultural eutrophication. As a result, highly polluted areas of water become dead zones that don't have enough oxygen to support a healthy aquatic ecosystem.

One of the most studied dead zones as a result of nutrient pollution occurs in the Gulf of Mexico. The Mississippi River flows into the gulf and brings with it nutrient runoff from agricultural fields in the Midwestern U.S. For a long time, scientists didn't understand what was causing huge numbers of fish to die off and whole regions of the Gulf of Mexico to display low levels of dissolved oxygen. After observing the ecosystem for a number of years, scientists realized that the worst dead zones occur after spring rains along the Mississippi River. These rains wash fertilizers and manure from agricultural fields into the river, which then carries these nutrients into the gulf. When these nutrients reach the gulf, the cycle of cultural eutrophication begins with an algae bloom and subsequent ecosystem disruption.

This pollution scenario isn't unique to the Gulf of Mexico. It's also common in other regions, such as the Mediterranean Sea and the Black Sea. In most cases, the dead zones are a recurring event; after the ecosystem recovers, a fresh flow of nutrients occurs and the cycle repeats itself.

Many cities have begun to monitor and limit the amount of nitrogen and other nutrients that enter nearby aquatic ecosystems in an effort to control nutrient pollution, oxygen sag, and cultural eutrophication. Another way to solve this type of water pollution problem is by finding ways to clean sewage and wastewater before it enters and damages an ecosystem; I describe this approach in the next section.

Breaking with Tradition: Wastewater Treatment and Constructed Wetlands

Rarely do you see raw, untreated human waste or sewage being dumped directly into water supplies in developed nations. Usually wastewater goes to a sewage treatment plant, where it's cleaned up before being released into nearby ecosystems (or reused as municipal water supply).

The sewage treatment plant cleans wastewater in three stages:

- **Primary treatment:** The first stage removes solid material and sediment from the water.

- **Secondary treatment:** The second stage uses bacteria to break down organic compounds and remaining biological matter. Air is pumped through the water tanks to encourage bacterial breakdown processes. Any remaining material sinks to the bottom of the tanks and is removed as sludge (and then burned or buried as landfill waste). The water then goes through a disinfection tank that uses chemicals or radiation to kill any pathogens remaining in the water.

- **Tertiary treatment:** The final stage removes the remaining nutrients (nitrogen and phosphorus) before releasing the water into the environment.

In traditional wastewater treatment, all three of these stages take place in tanks and lagoons at a water treatment facility. In recent decades, some urban areas have tried a new, more sustainable approach to treating sewage.

Some communities have used *constructed wetlands,* or manmade wetland systems and greenhouses, to treat wastewater in much the same way as traditional wastewater treatment systems but at a lower cost. Additionally, constructed wetlands offer an opportunity to create wetland habitat and natural space in a community.

Constructed wetlands don't change the primary wastewater treatment stage, but they do change the secondary and tertiary stages. Figure 16-2 illustrates the general stages of wastewater treatment using a constructed wetland. Basically, plants in the wetland remove nutrients and some metals from the water, while supporting organisms (such as bacteria) that break down organic waste matter.

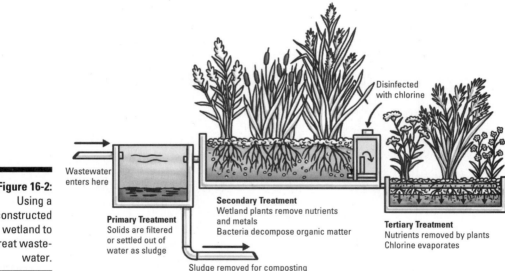

Figure 16-2: Using a constructed wetland to treat waste-water.

Wastewater enters here

Disinfected with chlorine

Primary Treatment
Solids are filtered or settled out of water as sludge

Secondary Treatment
Wetland plants remove nutrients and metals
Bacteria decompose organic matter

Tertiary Treatment
Nutrients removed by plants
Chlorine evaporates

Sludge removed for composting

Illustration by Lisa Reed

The advantages of using constructed wetlands include saving on energy costs by letting nature do the work and creating wetland habitats that support a healthy and diverse ecosystem. Not to mention, the wetlands are pleasant to visit while also transforming wastewater into clean water that's safe to release back into the environment.

Today scientists and wetland engineers are working on ways to make constructed wetland water treatment systems a reasonable solution to small-scale water treatment. Imagine treating wastewater from your home in a small wetland ecosystem in your backyard!

Chapter 17

What's Your Poison? Toxins and Infectious Disease

....

In This Chapter

▶ Knowing which toxins are most common in the environment

▶ Understanding how toxins accumulate inside you and move up the food chain

▶ Evaluating the risks toxins pose to human and environmental health

▶ Taking a closer look at modern infectious disease

....

Common sense tells you to steer clear of poisons because you know they can make you sick or lead to death. But did you know that poisons, or *toxins,* are all around you — in the environment, in your home, and in the products you use every day?

In this chapter, I describe the most common toxins that cause human health problems and environmental damage. I also explain where these toxins come from and how they get into the environment and your body. Then I define the different ways scientists and lawmakers decide which toxins are too risky to expose humans to and which ones are safe enough.

Because not all health problems are the result of toxins in the environment, I also spend time explaining a little about infectious diseases and describing how they develop resistance to the drugs scientists create to control them.

Identifying Common Toxins

Toxins are compounds that are extremely harmful even in small doses. In most cases, they interfere with basic cellular function, leading to death. *Toxicologists* (scientists who study toxins) classify toxic substances into the following categories:

✓ **Allergens:** An *allergen* is anything that triggers an organism's immune system to respond. In many cases, the allergens themselves aren't toxic, but they may trigger extreme responses by the immune system that lead to death. For example, some people with allergies to bee stings or peanuts may experience a swollen airway in response to these allergens; this swollen airway, in turn, restricts their breathing and may cause death. Formaldehyde used in building materials is a common allergen and results in sick building syndrome, which I describe in Chapter 15.

✓ **Carcinogens:** A *carcinogen* is a toxin that causes *cancer* (the out-of-control cell growth that creates tumors). Some carcinogens are also mutagens (see the later bullet point) because they damage the genetic information stored in a cell (though not all mutagens are carcinogens). Asbestos is one of the most common environmental carcinogens and is found in many building materials.

✓ **Endocrine disruptors:** Toxins that interfere with hormone function are called *endocrine disruptors.* The hormonal system of the human body, also called the *endocrine system,* regulates the activity and growth of certain organs. (Hormones lead most people to think about sex and reproductive development, but hormones manage a whole lot more than just that one system. Insulin, adrenaline, and endorphins are also types of hormones.) One type of endocrine disruptor, called an *environmental estrogen,* mimics the natural hormones in the body so closely that the body mistakes it for the real thing. When this occurs, the hormones don't send the signals they're responsible for sending, and as a result, the body's organs don't function properly.

✓ **Mutagens:** A *mutagen* is a toxin that damages the genetic material inside a cell, causing it to mutate or change. If these mutations occur in reproductive cells, they can cause birth defects and pass on to future generations. Arsenic, other metals, and some forms of radiation are mutagens.

✓ **Neurotoxins:** A *neurotoxin* is a chemical that interferes with an organism's nervous system. The nervous system moves messages to and from the organism's brain to run various bodily functions. When a neurotoxin disrupts this system, it can cause paralysis or brain damage. Metals such as mercury and lead are common environmental neurotoxins that can damage the brain and other important organs, such as the kidneys.

✓ **Teratogens:** Toxins classified as *teratogens* damage a fetus while it's developing in the mother's uterus. The amount of damage varies — from low birth weight to birth defects to death — depending on the type of teratogen and the degree of exposure. The most common teratogen is alcohol, which is why pregnant women are warned not to drink alcohol.

All toxins are hazardous. But not all hazardous materials (such as the ones I describe in Chapter 18) are toxins; some materials are dangerous because they're flammable or because they irritate without actually killing organisms.

Sticking Around: Persistent Organic Pollutants

The danger in some toxic substances is that they don't *biodegrade* or recycle back into the ecosystem. While many modern pesticides are made to degrade into their original elements and molecules so that they aren't toxic as they move through the ecosystem, other substances such as plastics and other manmade chemicals don't biodegrade. These nonbiodegradable compounds are called *persistent organic pollutants* (POPs) because they stick around for a long time, polluting the environment and causing illnesses. Some common POPs include

- ✔ **Bisphenol A (BPA):** You find BPA in polycarbonate plastics — the hard, clear plastics that many water bottles are made of. This compound is an endocrine disrupter and likely affects reproductive systems. Scientists are still studying the immediate health risks of BPA, but they have linked it to miscarriages and mental retardation in animals.

- ✔ **Polychlorinated biphenyls (PCBs):** You find PCBs throughout your home in the electrical wiring insulation, paint, flame retardants, and sealants (like the caulk around your bathtub). These toxins are manmade organic compounds that act as immune system depressants — keeping the immune system from properly responding to threats. Remember that in this case, *organic* simply means the molecules contain carbon.

- ✔ **Polybrominated diphenyl ethers (PBDE):** PBDE is a flame retardant commonly used in fabrics and furniture foam, although you also find it in plastics and appliances. It's chemically very similar to PCB and has been banned in Europe. In studies, it has been shown to disrupt thyroid function and cause neurological damage in laboratory animals.

- ✔ **Perfluorooctane sulfonate (PFOS):** You find PFOS in nonstick and stain-resistant products, including Teflon (for cooking pans), Gore-Tex (for weatherproof outdoor gear), and Scotchgard and Stainmaster (for carpets and upholstery). These compounds can withstand high levels of heat and are used in the manufacture of numerous products. They're carcinogens and have caused brain damage and reproductive damage in rats. Studies are currently underway to determine their effect on humans, and early results suggest that females may be more sensitive to their toxicity than males.

- ✔ **Phthalates:** You find phthalate compounds in everyday products such as makeup, plastic food packaging, body care products, and children's toys. Some phthalates have been shown to kill laboratory animals and have been linked to reproductive issues in humans. Studies are still underway to determine whether these endocrine disrupters are responsible for a trend toward lower sperm counts in men in the U.S. over the last 50 years.

- ✔ **Perchlorate:** Perchlorate is a chemical found in rocket fuel and other propellants. You may assume that you're safe from this toxin if you haven't been near a rocket recently, but perchlorate gets into water supplies and enters the food chain when farmers use that water to water animals and crops. It was widely spread across the U.S. during missile-testing activities through the mid-20th century. Perchlorate is an endocrine disrupter that scientists have found in recent human breast milk samples from across the U.S.

- ✔ **Atrazine:** Atrazine is one of the most commonly used herbicides in the U.S. Humans use it on crops, parks, golf courses, and other vegetation. Several European countries have banned atrazine, though it's still used in other countries around the world, including the U.S. It's an endocrine disruptor, and scientists find that nearly all surface waters in the U.S. have some amount of atrazine in them today.

Understanding the Ins and Outs of Toxic Exposure

You come into contact with toxins nearly everywhere: in the air you breathe, the water you drink, the food you eat, and the ground you walk on. These different *routes of exposure,* or ways of being exposed to toxins, present you with different doses of toxins. In this section, I describe how toxins accumulate inside of you and how they're magnified up through the food chain.

Accumulating in fatty tissue: Bioaccumulation

The main factor that determines how easily a toxin moves around the environment or around your body is its *solubility* (its ability to be dissolved). For example, some toxins are water soluble, meaning they dissolve easily in water and can go anywhere water goes. Other toxins are fat soluble, meaning they dissolve into other fats (oils) and then move through the environment or your body.

Toxins that are fat soluble are more likely to build up or accumulate over time inside your body. This *bioaccumulation* of toxins is dangerous to your health because what starts out as a very low and harmless dose of the toxins in your body gradually becomes a higher, more dangerous dose.

Magnifying up the food chain: Biomagnification

Some toxins are present in the environment in such small amounts that they're considered harmless if you're exposed to them. However, an interesting thing happens to these toxins as organisms ingest them and the toxins make their way through the ecosystem: What starts as a minor concentration of toxin in the environment becomes highly concentrated by the time it reaches the top predators in the food web. (See Chapter 6 for details on the food web and trophic levels.) This process is called *biomagnification*.

Environmental scientists look at a classic example of biomagnification in the way the pesticide DDT moves through the ecosystem. Figure 17-1 illustrates the biomagnification of DDT as organisms ingest it at increasingly higher trophic levels.

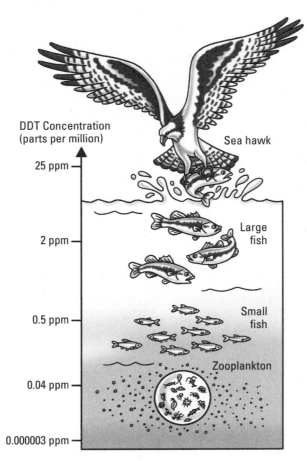

Figure 17-1: Bio-magnification of DDT.

Illustration by Lisa Reed

In the environment, DDT is present in low levels, often much less than 1 *part per million* or *ppm* in water (one molecule of toxin per one million molecules of water). These toxins are absorbed by the producers (algae) in the water, which are then eaten by the primary consumer, the zooplankton. When the zooplankton ingest the DDT, the toxic levels become concentrated to 0.04 ppm. Small fish then eat large numbers of the toxin-containing zooplankton, resulting in levels of 0.25 ppm in these fish. Larger fish then consume the contaminated small fish over the course of their lifetime, and the toxin magnifies to 2 ppm in each large fish.

The highest and most harmful doses of the toxin occur at the top of the food chain. Top predators, such as the fish-eating osprey (also known as the sea hawk) in the DDT example, have concentrations as high as 25 ppm. Why are the concentrations so high? Because these top predators have relatively long life spans and spend many years eating contaminated fish, allowing high levels of the toxin to build up in their systems.

Other toxins that scientists have observed biomagnifying throughout the food chain include mercury, arsenic, and PCBs. What seem like harmless amounts of these substances in the environment result in animal illness, death, and potential ecosystem disruption all because of biomagnification.

You may have heard warnings about eating too much tuna or other large fish because of dangerous levels of mercury. The danger is due to the way mercury biomagnifies through the aquatic food web. Today most mercury enters the environment from coal plant emissions and is present in low (nontoxic) levels. However, when organisms like fish consume this mercury, it accumulates in their tissue and sticks around. Each successive consumer in the food web accumulates a more concentrated amount of mercury. Top predators in aquatic ecosystems, such as tuna, swordfish, and sharks in the ocean and bass and trout in fresh water, have the highest concentration of mercury in their tissue — enough that humans (at the very top of the global food web), who consume large amounts of these fish, can become ill with mercury poisoning.

Mercury poisoning in humans may cause neurological damage, hypertension, increased sweating, rashes, shedding of skin, hair, and fingernails, and muscle weakness. Fortunately, you can reverse most symptoms of mercury poisoning by reducing the levels of mercury in your body. However, exposure to high levels of mercury in childhood or in utero may result in permanent health issues.

Risky Business: Assessing the Dangers of Chemicals

In setting up regulations about the use of toxic chemicals and the degree to which toxins are allowed to enter the environment, policymakers ask scientists to perform risk assessment. *Risk assessment* helps determine how dangerous a particular toxin is to human and environmental health. In this section, I describe a few ways of assessing risk through scientific study, and I explain how perceived risk is different from actual risk. I end the section by explaining what policymakers do after the risk assessment stage — pick a risk management strategy.

Measuring risk

To inform policymakers and resource managers about the dangers of toxins in the environment, scientists set up studies that help measure or quantify potential risk. One approach to quantifying risk is calculating measurable values of *probability,* or how likely something is to occur. For example, scientists calculate toxic risk as the likelihood of being exposed to a toxin multiplied by the likelihood of becoming ill after being exposed to that toxin.

To perform a risk assessment, scientists have to collect large amounts of data from experiments and studies. The three types of studies that scientists use to collect data about toxins are

- ✔ **Dose-response studies:** In *dose-response studies,* scientists in a laboratory alter the amount of chemical exposure, often using lab animals such as mice, and observe the results. As the dose of the potential toxin increases, the animal may become ill or die. Scientists record what dose of the toxin resulted in what type of reaction and use that information to estimate how a similar dose would affect a human.

- ✔ **Retrospective studies:** *Retrospective studies* are one way to use information from human subjects to understand how a toxin affects human health. In a retrospective study, scientists collect information from people who have already been exposed to a chemical at some point in the past. They monitor the exposed individuals, as well as a control group of individuals who haven't been exposed (see Chapter 2 for details on setting up experiments and using control groups). This type of monitoring and retrospective study may continue for several decades as scientists collect as much information as they can about the toxic effects.

> ✔ **Prospective studies:** *Prospective studies* are the opposite of retrospective studies in that they look toward the future rather than back at the past. Scientists set up a prospective study by identifying a group of people that may become exposed to a toxic chemical in the future and tracking their habits for a number of decades. These studies are more difficult to control than the other two types because people's habits (such as what and how they eat and whether or not they smoke tobacco) may change over time or may not be reported correctly.

No matter which type of study scientists use to assess the risk of toxins, they must also consider the possible effect of multiple risks adding together. For example, although one risk may be dangerous and another may be somewhat less dangerous, when they combine, the effect may be extremely dangerous.

When the combined effect of two risks is greater than expected, that combination is called a *synergistic interaction*. A common example of a synergistic interaction is the risk of cancer from being exposed to asbestos and from smoking cigarettes. Both cigarettes and asbestos are carcinogens and have a risk of causing cancer. However, a person who has been exposed to asbestos and who smokes cigarettes is at much greater risk of developing cancer than someone who has been exposed to one or the other.

Perceiving risk

While some risk can be measured through scientific experiments and studies, people's perceptions of risk may not be based on hard data. For example, many people perceive flying in airplanes as being an extremely risky activity, considering they could die if the plane were to crash. In truth, though, the likelihood of dying in a plane crash is fairly low when compared to other risks that lead to death, such as smoking cigarettes or driving a car.

In this way, perceived risk isn't always rational or based on reason; instead, it's based on perceptions and emotions. In most cases, the less familiar you are with something, the riskier you perceive it to be. Figure 17-2 illustrates the relationship between people's fear or perception of risk and their familiarity with a specific technology or science.

The information in Figure 17-2, collected and published by Paul Slovic in a 1987 article in the journal *Science,* illustrates that perceived risk isn't always rational. If you compare the results on the graph (particularly the items listed in the lower-left corner as low risk) with the list in Table 17-1 of leading causes of death in the U.S., you can see that while the public perceives smoking to be relatively low risk, it's actually the third-leading cause of death.

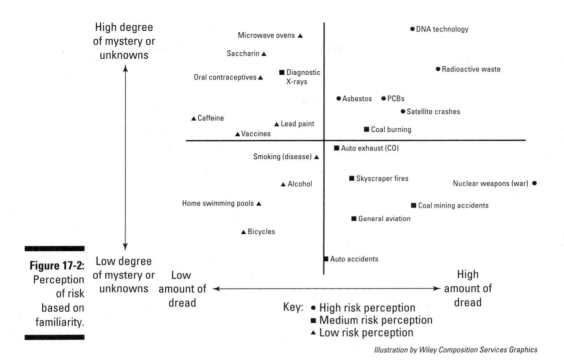

Figure 17-2:
Perception
of risk
based on
familiarity.

Illustration by Wiley Composition Services Graphics

Table 17-1	Leading Causes of Death in the United States
Cause	*Odds*
Heart disease	1 in 2
Cancer	1 in 3
Smoking	1 in 4
Lung disease	1 in 15
Pneumonia	1 in 30
Car accident	1 in 100
Fire	1 in 1,000
Plane crash	1 in 5,000
Drowning	1 in 10,000

Choosing a risk management strategy

Obviously, you can't avoid all the risks out there, so you need to manage them according to how risky they are. In the case of toxins in the environment, policymakers use risk assessment to help them manage the risk. Ideally, policymakers would call for the removal of any chemical that is at all risky to human health so that no one would be exposed. But in reality, public health risks are pitted against chemical industries, which view risk management very differently from most individuals. Here's how the two approaches to risk management vary:

- **Presuming innocence until proven guilty:** Some industries and policymakers view the potential risk of chemicals from an *innocent-until-proven-guilty perspective.* Risk managers and policymakers with this view believe that the government shouldn't regulate or restrict chemicals until scientific data prove that they're definitely harmful. This approach allows products to reach the market (and therefore the consumer) more quickly. However, it also means that by the time scientific data determine a chemical to be too dangerous for public health, much of the public has already been exposed.

- **Applying the precautionary principle:** Risk managers and policymakers who practice the *precautionary principle* think that reducing potential health hazards before exposing people to a possible toxin is worth the time and money. This approach requires that industries do more extensive testing on products before they release them to the public and may mean that some products are never released.

Tracking Infectious Disease

Toxins aren't the only things that affect the health and welfare of people around the world. Many health problems are the result of infectious diseases, which are caused by *pathogens* (commonly called *germs*) such as bacteria, viruses, and other microorganisms. In this section, I describe some modern pathogens, and I explain how the use of antibiotics to fight off infectious disease leads to stronger, harder-to-kill pathogens.

Emerging on the scene: Modern infectious disease

Many diseases caused by pathogens have been around for a long, long time. For example, the bubonic plague, a disease caused by bacteria, killed millions of people in Europe during the 1300s. Malaria and tuberculosis are other examples of pathogenic diseases that have been around for many centuries and are still causing illness today.

Some infectious diseases are relatively new to the disease scene. These diseases, which have either recently been described for the first time or haven't been common for the last few decades, are called *emergent infectious diseases*. Examples of emergent diseases recognized today include the following:

- **AIDS/HIV:** AIDS stands for *acquired immune deficiency syndrome* and is caused by a virus called the *human immunodeficiency virus* (HIV). Although AIDS is well-known today, it was an emergent disease in the 1970s and 1980s. Researchers didn't know how it made its way into the human population until 2006 when they found the virus in an African chimpanzee population and realized that local hunters were exposed to the virus by eating these animals. Due to the biological similarity between chimps and humans, the virus easily transferred into the human population and began to spread. In recent decades, scientists have developed strong antiviral medications that effectively reduce viral infection and the resulting AIDS illness, but they're still very expensive to produce and not yet in widespread use.

- **Ebola fever:** *Ebola fever* is caused by the *Ebola virus,* named after the river in Africa near where scientists first discovered the virus in humans. After being contracted, the virus causes severe vomiting, fever, and usually death after only two weeks of illness. Scientists have not yet figured out the source of this virus or found methods for fighting it.

- **Mad cow disease:** Also known as *bovine spongiform encephalopathy, mad cow disease* first appeared in the 1980s as a brain disease in cows. Unlike other pathogenic diseases, mad cow is caused by a protein that mutates into a pathogen, making it more difficult to control through the usual methods. After scientists recognized that a cow or person had to eat an infected cow to contract the disease, government regulators and the beef industry took steps to make sure cattle didn't feed on infected animal remains. Now the risk of passing on mad cow disease to humans is significantly reduced.

- **West Nile virus:** *West Nile virus* is carried by mosquitoes and transmitted through mosquito bites. It has been common in Africa since the 1930s, but it didn't appear in North America until 1999. Since then, the virus has spread across the entire continent. It affects birds, humans, and more than 200 other species. Humans that are infected by West Nile experience brain swelling that may lead to death. In an effort to control the spread of West Nile virus, some regions apply pesticides to reduce mosquito populations.

✔ **Zoonotic influenza:** *Influenza,* or *flu,* viruses common in other species have infected humans more than once throughout history. (*Zoonotic* simply means that the disease moves from an animal species to humans.) The 1918 outbreak of Spanish bird flu that killed millions of people around the world is one such instance. More recently is the outbreak of the H1N1 flu virus common to pigs and birds (and related to the earlier Spanish flu virus). Although humans are exposed to many flu viruses and build immunity to them, when a new strain jumps from one species to another, the newly exposed species (in this case, humans) isn't prepared to fight it off. As a result, many humans have suffered or died from viruses like H1N1.

Evolving resistance

Scientists have worked hard to develop medications to battle infection. Unfortunately, in many cases, the drugs they develop are effective only for a short time before the pathogens that cause infection become immune to the drugs and scientists have to develop stronger versions. Battling pathogens presents this problem of drug resistance because of natural selection and evolution.

In any given population of organisms, only those with the most adaptive and successful traits survive to reproduce. So when a group of pathogenic organisms is exposed to an antibiotic drug, the drug destroys most of the pathogens, but a few may survive because of *gene mutations* (random genetic changes) that make them somehow resistant to the drug's effects. As a result, the next generation of pathogens, which stems from these resistant organisms, is also resistant to the drug. The antibiotic is no longer useful to control this resistant generation of pathogens, and scientists have to create a new antibiotic that's effective at stopping these stronger, mutated pathogens. (I talk more about genetic evolution and natural selection in Chapter 12.)

The ability of pathogens and other bacteria to develop resistance to antibiotics is one reason why you shouldn't take antibiotics for illnesses that are viral. Viral illnesses are caused by viruses and require anti-viral drugs (which, although rare, are currently being developed). Only diseases caused by infectious bacteria can be treated with antibiotics. By taking antibiotics when you don't need them (like when you have a viral illness) or not taking the recommended dosage that your doctor prescribes, you help pathogens build up drug resistance so that antibiotics may not be as effective the next time you do need them.

Cleaning your house the nontoxic way

Some of the most common toxic products in your house are the chemicals you use to clean it. In fact, many of these products have labels that express how dangerous the chemicals can be:

✔ **Caution:** Mildly toxic

✔ **Warning:** Moderately toxic

✔ **Danger:** Highly toxic or otherwise hazardous (such as flammable)

✔ **Poison:** Extremely toxic

To reduce the presence of these toxins in your home, you can use safe, nontoxic, and biodegradable cleaning products instead. Here are just some of the effective yet safe cleaning supplies you likely already have on hand:

✔ **Baking soda:** Scours and deodorizes

✔ **Lemons:** Contain a strong acid that kills bacteria

✔ **White vinegar:** Cuts through grease and wax buildup and removes mildew

✔ **Borax (sodium borate):** Cleans, deodorizes, and disinfects

✔ **Cornstarch:** Cleans windows and carpets and polishes furniture

With a little household chemistry, you can clean almost everything in your house — without the use of toxic chemicals. Here are a few nontoxic cleaning tips:

✔ Sprinkle baking soda on a surface and then gently scrub it with a sponge. If the grime is extra thick, add some salt to the baking soda on the surface.

✔ Mix 2 tablespoons of white vinegar (or lemon juice) with a gallon of water and use it in a spray bottle to clean windows.

✔ To clean up liquid spills on carpet, cover the area with cornmeal, let it sit and absorb, and then vacuum it up.

✔ Use baking soda or cornstarch to deodorize carpets. Sprinkle about a cup across a medium-sized room, wait 30 minutes, and then vacuum.

✔ Spray a solution of vinegar and warm water (1/4 cup of vinegar and 30 ounces of warm water) onto a cloth or cotton mop and use it to clean hardwood floors.

✔ Clean your oven with a paste of baking soda and water. Apply the paste to the inside of your oven and let it sit overnight. Then get busy scrubbing the grime away.

Chapter 18

A River of Garbage Runs through It: Solid and Hazardous Waste

Some types of waste can break down and be reused by other organisms, thus being recycled through the ecosystem. In the natural environment, what appears as waste from one organism is food for another (think decomposers; see Chapter 6). Humans, however, are the only creatures on Earth that generate garbage and solid waste that can't recycle through the environment and instead must be safely stored or reused. In this chapter, I focus on the materials humans create, use, and throw away.

Garbage and solid waste disposal creates problems that spill into many other issues environmental scientists want to solve; it creates air and water pollution, uses land space, and can release toxins into the environment. As I explain in Chapter 3, matter is neither created nor destroyed, so everything you throw away must go somewhere — and you may be surprised at where it ends up.

In this chapter, I describe how garbage is created and where it goes after you throw it out. I also cover hazardous waste and solutions for cleaning it up.

Wading into the Waste Stream

Waste is created at many different stages of a product's life cycle, beginning with raw materials and ending when a consumer throws the product in the garbage can. The term *waste stream* describes the ongoing creation of waste in the production, use, and disposal of material items. Figure 18-1 illustrates the cycle of waste creation and disposal, including how some things are recycled in the waste stream.

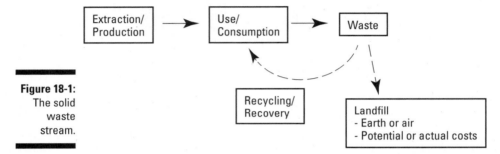

Figure 18-1:
The solid
waste
stream.

Illustration by Wiley, Composition Services Graphics

As you can see in Figure 18-1, the first stage of waste generation occurs when raw materials are extracted from the environment. The manufacturing process of the goods produces additional waste, and then the product itself becomes waste when it's no longer useful to the consumer.

Waste is categorized according to how and when it's generated along the waste stream. Generally speaking, you can sort nonhazardous waste into three broad categories (I cover hazardous waste later in this chapter):

- ✔ **Industrial waste:** *Industrial waste* occurs during the resource extraction (such as mining) and production or manufacturing stages.

- ✔ **Agricultural waste:** *Agricultural waste* results from farming and animal production.

- ✔ **Municipal waste:** By far, the largest source of solid waste is *municipal solid waste* (MSW), which includes waste collected from households, businesses, schools, hospitals, and other buildings in towns and urban areas. Anything collected by your local waste or recycling company is municipal solid waste.

Although other industrialized nations also produce large amounts of waste, the U.S. is by far the leader of the modern throw-away society. MSW accounts for the largest amount of waste in the U.S., with more than 200 million tons a year. In the name of convenience, many products are manufactured specifically to be used and then thrown away. Other countries, striving to compete in the world market and seeking the prosperity seen in more industrialized nations, are quickly increasing their levels of waste. Because these developing nations produce and manufacture most of the world's goods, they often create more industrial waste as well and are left to find solutions for the disposal of large amounts of waste.

Disposing of Waste

After you're done using a product and throw it away, it has to go somewhere because it's composed of matter and doesn't just disappear. Some products, such as paper, aluminum, and some plastics, are commonly recycled (I cover recycling in the next section), but more than half of American waste is left to be buried, burned, or dumped somewhere away from your home, neighborhood, or city. Here are some of the most common disposal methods.

Tossing it aside and covering it up: Dumps and landfills

Although many people use the terms *dump* and *landfill* interchangeably, these disposal methods aren't quite the same. *Dumps* are open-air sites where people leave garbage, old appliances, and anything else that's no longer useful. Dumps as a means of waste disposal are most common in developing nations, where they're likely to allow contaminants into the soil and pollute nearby water and air.

Landfills are one of the most common places for municipal waste disposal. A *landfill* is different from a dump because the waste in a landfill is covered up, although both types of sites have traditionally created similar environmental hazards. Today, most U.S. landfills are *sanitary landfills,* designed and engineered to reduce ecosystem contamination while storing solid waste. Figure 18-2 illustrates the following important features of a sanitary landfill:

- ✓ Liner and cap made of clay or manmade material to keep pollutants from entering the soil and contaminating groundwater

- ✓ Drainage layer where *leachate* (contaminated runoff water) is collected and sent to the surface for cleaning and treatment

- ✓ Rainwater pond to keep rainwater from seeping through the landfill and creating more contaminated leachate

- ✓ Methane extraction system to remove methane gas produced by the waste (This methane can be used as an energy source; see Chapter 14 for details.)

> ✔ Monitoring systems to check levels of contaminants and levels of
> methane gas in the ground surrounding the landfill
>
> ✔ Cover of sand, clay, and soil, planted with vegetation

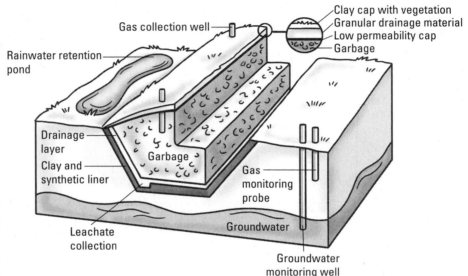

Figure 18-2:
Features of
a sanitary
landfill.

Illustration by Lisa Reed

Sanitary landfills can only protect the environment from so much. Although
their features are constructed to protect the ecosystem from everyday garbage
contaminants, these sites aren't appropriate for disposing of hazardous or
toxic waste. Flip to the later section "Handling Hazardous Waste" for info on
dealing with this kind of waste.

Burning trash: Incineration

Another commonly used method of waste disposal is *incineration,* or burning.
Not all waste can be incinerated, but garbage composed of paper, plastic,
food, and yard waste is relatively safe and easy to burn. The result of
burning waste is *ash,* a term that includes any nonorganic material that
remains (in other words, the stuff that won't burn).

Incinerators can also generate energy through a new technology called
energy recovery. An *energy recovery* system, sometimes called a *waste-to-
energy* system, uses the heat energy released by burning waste to generate

electricity or steam. Although this technology is a positive outcome from incineration, burning trash still isn't a perfect solution. Trash incinerators are expensive to build, and they contribute to environmental problems such as air pollution. In particular, incinerating toxic materials (such as batteries) or metals can release pollutants into the atmosphere (see Chapter 15).

Exporting electronics: E-waste disposal

In the modern era, much of the waste that developed nations generate is in the form of broken or outdated electronics. The U.S. alone throws away almost 50 million computers every year! This *electronic waste,* or *e-waste,* often contains toxic materials such as metals and plastics.

These materials can't safely be disposed of in landfills or incinerators, so disposing of them consists of shipping them overseas to developing nations under the guise of recycling. In this case, *recycling* means scrap recovery; folks in the developing countries remove and sell internal components such as copper wiring and bits of precious metals, leaving what remains to pollute the environment.

Exporting e-waste to other countries has been illegal in the U.S. since 1989, yet somehow it continues to occur. Most U.S. e-waste is purchased by so-called recyclers in China, who deposit it in large dump sites where the poorest people, including children, spend their days amid the toxic piles trying to recover anything they can resell. Sometimes the toxic materials are burned to release desired metals, even though the reason the U.S. exported these materials in the first place is that they're too toxic to incinerate safely. Regardless of the recycling method, the waste often ends up contaminating water supplies. Countries in Africa, India, and Eastern Europe receive and recycle the e-waste from Western Europe under similar conditions.

Swimming in Waste: The Ocean Garbage Patches

More garbage than you may realize is simply thrown into the world's oceans. Bottles, cans, plastic bags, and other packaging are the most common kinds of waste in the ocean, followed by waste from the fishing industry, such as large nets and other gear. The presence of garbage in the oceans disrupts the ecosystem and endangers animal life by entangling or strangling sea creatures, for example. Disposing of solid waste in the world's oceans is reaching epic and ecologically dangerous proportions.

Swirling and whirling: Concentrating trash in ocean gyres

As ocean water circulates around the globe, it creates large, circular patterns of current rotation called *gyres* (more on these in Chapter 7). These gyres are driven by wind blowing across the surface. Figure 18-3 illustrates the locations of the five major ocean gyres.

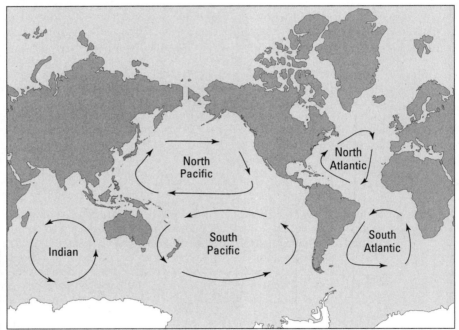

Figure 18-3:
The five major ocean gyres.

Illustration by Wiley, Composition Services Graphics

Ocean currents carry and concentrate garbage in the center of these gyres, creating large patches of floating trash. Each gyre has an ocean garbage patch; the largest is the Great Pacific Garbage Patch in the North Pacific Ocean.

Figure 18-4 is a photograph of a small sample of trash picked up in a net out in the Pacific gyre. It includes large pieces of Styrofoam, rope, plastic bottles, and other plastic bits — even an old toilet seat!

Figure 18-4:
Sample of
trash
collected
from the
Pacific gyre.

Photograph by UIG via Getty Images

The most damaging result of ocean garbage is how it affects the ecosystem. Examining the contents of seabird stomachs helps demonstrate the effects of this sea of trash on marine ecosystems. Scientists increasingly find items such as plastic lighters, bottle caps, toy miniatures, and other random bits of plastic in the stomachs of dead albatross chicks (the *albatross* is a scavenging seabird); the albatross parents mistook these discarded bits as food floating on the ocean surface and fed them to their hatchlings. One animal's stomach reportedly contained 1,603 pieces of plastic.

Fighting photodegradable flotsam

Although the idea that ingesting lighters and other plastics rather than food can endanger or kill marine organisms may seem obvious, what's less obvious is the danger posed by even small bits of plastic that cloud up ocean water in the gyres.

Most of the garbage in the oceans isn't simply floating on the surface. Rather, much of the trash is suspended beneath the surface in the water column; in particular, I'm talking about *photodegradable* plastics, which break down into smaller particles when exposed to sunlight.

Because they're manmade, plastics don't break down into organic molecules or their original elemental atoms and can't be recycled into the ecosystem. Instead, they break down into microscopic pieces of plastic that create the following problems:

✔ Microscopic plastic takes up space that marine primary producers (algae) previously occupied, disrupting the food chain.

✔ Organisms mistake tiny plastic bits for food and ingest them, allowing manmade toxins such as bisphenol-A (see Chapter 17) to enter the marine food chain.

✔ Bits of plastic provide a surface that bacteria can inhabit, fostering colonies of bacteria and causing illness when ingested by other organisms.

✔ High concentrations of microscopic plastics block sunlight from reaching organisms deeper in the water.

Consider that every piece of plastic ever made still exists today and that many of those plastics are known to contain toxins (check out Chapter 17 for details on different types of toxins in plastic). By disposing of these plastics in the world's oceans, humans have ensured that these materials will enter, and forever cycle through, the global ecosystem's food chain.

Stopping trash at the source

In the last few decades, scientists have only begun to understand how the trash in the ocean garbage patches is affecting the larger marine ecosystem. What many have realized is that cleaning up the oceans will take international cooperation and a commitment to caring for and sharing ocean resources.

The best way to tackle the problem of ocean garbage pollution is to start at the source — reduce consumption of plastic goods, reuse what you've already purchased, and recycle what is no longer useful. Reducing reliance on plastic goods is the best way to keep them from ending up in the oceans.

Filling in the facts

As you think about your own contribution to the waste stream, consider the following facts collected by the Clean Air Council:

✔ The average office worker uses 500 disposable cups every year — more than one per day!

✔ Almost 4 million tons of office paper are used every year in the U.S.

✔ Plastic bags are the second most common plastic item collected from ocean garbage patches.

✔ In 2003, 290 million old tires were added to landfills or burned.

✔ Enough gift ribbon is thrown away each year to wrap a bow around the entire world.

✔ 30 percent of municipal waste is made up of containers and packaging.

✔ The average American citizen creates 5 pounds of waste each day.

Remember: Reduce, reuse, recycle, and compost to decrease your waste stream.

Reduce, Reuse, Recycle (and Compost): Shrinking the Waste Stream

The solution to managing and ultimately shrinking the waste stream lies in the three *R*s (reduce, reuse, and recycle) and one *C* (compost). It really is as simple as it sounds. By asking yourself whether you really need a product (reduce), whether you can use that old product again (reuse), or whether the materials can be used another way (recycle or compost), you go a long way in changing how much waste you produce. Here's how it works:

✔ **Reducing:** The first step in reducing the amount of solid waste for disposal is to decrease the amount of materials introduced into the waste stream. This process means *source reduction,* or conserving resources in the manufacture of goods. For example, some products today advertise that they use less plastic or less overall packaging than they used to; this method of source reduction reduces the amount of materials destined to end up in a landfill.

✔ **Reusing:** Reuse products (for example, mailing envelopes and shipping boxes) as long as they're useful instead of throwing them away immediately. In some cases, reusing an item may require repairing it, but the cost of repair (including labor and energy) may well be worth keeping that item out of a landfill for a few more years.

Just because you have no further use for a product doesn't mean that someone else doesn't. One way to reuse useable goods is to donate them to a thrift store or community collection center that redistributes such items to people who can use them.

✔ **Recycling:** *Recycling* means that the waste materials get sent back to the beginning of the waste stream and are used to produce new products. Manufacturers are increasingly using recycled plastics to make bottles, bags, and even clothing; recycled cardboard, paper, and packaging are also common. Recycling one material (such as plastic bottles) into a new material (such as fleece jackets) is called *open-loop recycling*. Recycling an object into a new version of that same object (such as creating new aluminum cans out of recycled old cans) is called *closed-loop recycling*.

✔ **Composting:** Another form of recycling or reusing that reduces the solid waste in landfills and incinerators is *composting*. *Compost* is a mix of decaying organic matter that can provide nutrients for healthy soil. You can compost items such as eggshells, banana peels, coffee grounds, grass clippings, leaves, and even paper napkins and pizza boxes (which can't be commercially recycled because they contain food residue) and transform them into useful organic matter.

Some cities have developed municipal composting facilities to complement recycling and waste disposal programs. Composting waste requires that you separate the organic and truly biodegradable materials from your trash and place them in a separate bin for collection. The waste then goes to a compost facility, where it's added to everyone else's organic waste and transformed into useful garden compost that can be sold in bags at a local nursery.

Composting also works on a smaller scale with kitchen composters, worm bins, or backyard composters. Check out *Composting For Dummies* by Cathy Cromell and the National Gardening Association (Wiley) and www.howtocompost.org for details on composting and on what type of household composting system may work for you.

Handling Hazardous Waste

Many types of solid waste damage the environment, disrupt ecosystems, and can affect human health. However, most municipal solid waste isn't considered hazardous. *Hazardous waste* is a special category for waste material that is especially harmful to human health and ecosystems.

Hazardous waste disposal is regulated by laws that define hazardous waste as any solid, liquid, gas, or sludge that is

- ✔ Fatal to humans or lab animals in small doses

- ✔ Toxic to humans or other organisms

- ✔ Corrosive (like battery acid)

- ✔ Explosive or highly reactive (or volatile)

Most hazardous waste comes from chemical and fossil fuel industries and mining operations. The materials are often treated and recycled or stored at the production site. Manufacturers must follow careful procedures of treatment before disposing of hazardous waste in landfills or other storage. Although methods exist for disposing of hazardous waste relatively safely, the best choice is really to reduce hazardous waste production at the source.

Cleaning up the mess: The Superfund

In the 1980s, the U.S. passed laws to regulate the production, treatment, and disposal of hazardous waste. One of these laws, the *Comprehensive Environmental Response, Compensation, and Liability Act* (CERCLA) established a source of funding to clean up sites contaminated with hazardous waste. The goal of CERCLA (also called the *Superfund Act*) is to react quickly to contain and clean up abandoned hazardous waste sites that pose serious and immediate danger. The act also gives the Environmental Protection Agency (EPA) the right to sue responsible parties whose actions (for example, failing to properly contain or dispose of hazardous waste) it deems are a danger to the environment.

The Superfund created by CERCLA is a pool of money that was originally generated by taxes on industries that create hazardous waste but is now largely funded by public taxes. This money is used to cover the enormous expense of cleaning up hazardous sites listed on the *National Priority List* (NPL).

A site qualifies for the NPL if it's considered especially dangerous to human health. The substances most commonly associated with NPL sites include the following:

- ✔ Lead
- ✔ Trichloroethylene
- ✔ Toluene
- ✔ Benzene
- ✔ PCBs
- ✔ Chloroform

✔ Phenol

✔ Arsenic

✔ Cadmium

✔ Chromium

Most NPL sites are abandoned industrial facilities or old garbage and chemical dumps. Groundwater, surface water, and soils surrounding these sites are contaminated and create serious health dangers to people who live in the area.

Letting microbes do the dirty work

One of the most innovative ways to tackle environmental damage by toxins and pollution is to let nature do the dirty work of cleaning up the mess. Specifically, certain *microbes,* or microscopic organisms that live in soil and water, can help with *bioremediation,* the natural cleanup of contaminants in the environment, by eating chemical compounds that would otherwise harm the environment. For example, some microbes eat oil and have been useful in cleaning up the oil spilled in ocean waters. When oil-eating microbes digest the oil, they transform it into harmless waste products such as water and carbon dioxide.

The trick to bioremediation is providing the microbes with the proper conditions for maximum growth and function. If they don't have proper conditions of temperature, water, and nutrients (fertilizers), they may not be able to digest the contaminants properly or effectively. For this reason, scientists who use microbes for bioremediation try to provide the hard-working organisms with everything they need to get the job done.

One of the downsides to bioremediation is the addition of nitrogen- and phosphorus-based fertilizers to the ecosystem. Another is that not all contaminated sites are easy to access with microbes. For example, in some areas, using techniques of bioremediation to clean up groundwater requires a system of drilled wells and pumps to circulate microbes, nutrients, and air into the contaminated water deep underground. Finally, using microbes to clean up pollutants sometimes takes months or years (depending on where the site is and how many contaminants are present), which can be frustrating for people who want more immediate results.

Generally speaking, however, using bioremediation is a cost-effective and environmentally friendly approach to cleaning up certain pollutants in water or soils. In fact, the Environmental Protection Agency employs bioremediation at more than 50 Superfund cleanup sites across the nation.

Danger lying dormant: Bringing up brownfields

Although the Superfund takes care of hazardous sites posing the biggest threats to public health and environmental safety, many contaminated sites aren't "bad enough" to be included on the NPL. Abandoned industrial or commercial sites that are potentially polluted and remain undeveloped are called *brownfields*.

The degree of pollution at many brownfields is unknown. Developers are wary of being held liable for the cleanup costs, so the sites remain abandoned and ignored, left to contaminate the environment with any pollutants or toxins that remain from previous industrial, commercial, or business use. Common brownfields include old gas stations or dry cleaning businesses, as well as larger industrial sites.

Chapter 19

Is It Getting Warm in Here? Modern Climate Change

*I*n the early 20th century, scientists began asking questions about a warming trend they observed in climate data. At the time, it seemed almost impossible that anything humans were doing could significantly affect a system as large and complicated as the Earth's climate. In the last 50 years, however, as scientists have gathered and analyzed more and more data, they've concluded that, indeed, the Earth is getting warmer and humans are a real force of nature, rivaling volcanoes in their ability to alter the chemistry of Earth's atmosphere.

Environmental scientists called *climatologists* seek to understand the causes of global-scale changes in the climate system. They look into the past, observe the present, and try to predict the future based on their understanding of how the climate system functions.

In this chapter, I provide a basic explanation of the observed changes to Earth's climate, as well as scientists' current conclusion: that human activities are contributing to the observed changes in Earth's climate system. I begin by explaining how greenhouse gases in the atmosphere function to keep the Earth's surface warm and how increasing levels of these gases lead to warmer temperatures across the globe. I describe the most common greenhouse gases and explain where they're produced, both by natural processes and by human activities. I also explain what scientists know about how climate varies through time and why they have concluded that what they observe today is a result of human actions over the last 100 years. At the end of the chapter, I describe current and future effects of a changed climate system on the environment and human society.

Hothouse Humans: The Greenhouse Effect

Life on Earth couldn't exist without the protective cover of gas created by the Earth's atmosphere. Millions of years ago, gas molecules emitted from volcanoes began filling the atmosphere. These gases, along with oxygen added by photosynthetic organisms, began forming a blanket of protection from the intense radiation from the sun. Only after this layer of protective gas was in place did organisms on Earth's surface thrive and evolve into the complex life forms of today. In fact, all living things today, including humans, are adapted to survive in this hothouse called Earth.

In the upper atmosphere, a layer of protective ozone molecules blocks the most dangerous radiation from the sun (see Chapter 15 for details on the ozone layer). Other gases found in Earth's lower atmosphere are called *greenhouse gases* (GHGs). These gases make up only a small percentage of the overall gas content of the atmosphere and help keep temperatures near the surface of the Earth relatively comfortable.

Figure 19-1 illustrates how the natural greenhouse gases keep the Earth warm — but not too warm. This process is called the *greenhouse effect*.

The Greenhouse Effect

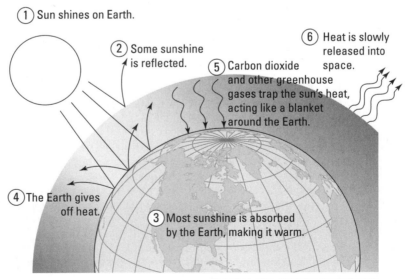

① Sun shines on Earth.

② Some sunshine is reflected.

⑥ Heat is slowly released into space.

⑤ Carbon dioxide and other greenhouse gases trap the sun's heat, acting like a blanket around the Earth.

④ The Earth gives off heat.

③ Most sunshine is absorbed by the Earth, making it warm.

Figure 19-1: The greenhouse effect.

Illustration by Wiley, Composition Services Graphics

Anytime GHGs are added to the atmosphere, this layer becomes thicker and more effective at trapping heat close to the Earth's surface. In Chapter 7, I describe how ocean currents and atmospheric circulation move heat energy around the planet, creating patterns of weather and ecosystems on Earth. As more and more GHGs are added to the atmosphere (and the greenhouse effect intensifies), these global circulation patterns work to distribute the warmth, resulting in warmer air and ocean temperatures all over the world.

Filling the air: The top six greenhouse gases

Scientists studying climate change recognize that greenhouse gases are added to the atmosphere through both natural processes and human activities. Here's a quick look at how the major GHGs are produced:

- **Water vapor (H_2O):** Water vapor occurs naturally as a part of Earth's hydrologic cycle (see Chapter 6). Water vapor levels vary in relation to temperature: More water evaporates into the atmosphere in regions of warmer temperatures.

- **Carbon dioxide (CO_2):** Carbon dioxide is naturally emitted into the atmosphere through volcanic eruptions. Humans add carbon dioxide to the atmosphere when they burn fossil fuels (coal, oil, and natural gas) for energy and transportation.

- **Methane (CH_4):** Methane is created naturally as organic matter decomposes in environments without oxygen, such as wetlands or the intestines of termites and cows. It's also a major component in natural gas, which humans use for energy production. Methane enters the atmosphere from garbage landfills as household waste decomposes and from cattle feedlots where cow digestion produces methane.

- **Nitrous oxide (N_2O):** Nitrous oxide occurs naturally as part of the nitrogen cycle when nitrates in the soil or oceans are converted to gas form (see Chapter 6). Humans add nitrous oxide to the atmosphere when they add nitrogen-rich fertilizers to their crops and the excess nitrates transform into gas.

- **Ozone (O_3):** Ozone occurs naturally in the upper levels of the atmosphere (called the *ozone layer*), but in the lower atmosphere, molecules of ozone act as greenhouse gases when they're created by photochemical reactions with air pollutants (see Chapter 15).

✔ **Chlorofluorocarbon (CFC):** Chlorofluorocarbon is a manmade molecule that didn't exist in the atmosphere until human industry began producing it in the early 20th century. Environmental laws now heavily regulate the use of CFCs because they're linked to the destruction of the ozone layer and they function as a GHG, trapping heat near Earth's surface (see Chapter 15 for more details on CFCs and the ozone layer).

Although all these gases contribute to warming the Earth, the greatest increase in GHGs in the atmosphere over the last century is a direct result of human activities that have sped up the naturally slow rate of carbon recycling through the Earth's system.

Recognizing human influences

Humans have played a role in Earth's climate system since the dawn of the industrial age, when they first began to use fossil fuels for the manufacture of goods and transportation. Figure 19-2 illustrates the increasing amount of carbon dioxide in the atmosphere over the last few hundred years.

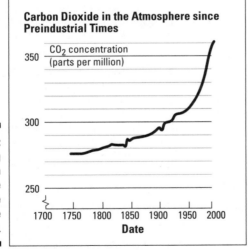

Figure 19-2:
Increasing carbon dioxide in the atmosphere since 1750.

Illustration by Wiley, Composition Services Graphics

To really grasp the effects human industrial activities have on the Earth's climate, you need to remember that Earth is a closed system, in which matter doesn't enter or leave and is instead continually recycled through the system (see Chapter 6 for more details). Understanding that Earth is a closed system may help you recognize that the outputs of human industrial activity influence the climate in the very same ways that natural GHG outputs do — by increasing temperatures. As human populations have grown and human civilizations have spread and industrialized, the human influence on the climate system has also grown.

The carbon that forms fossil fuels has been buried deep in the Earth for millions of years and, through geologic processes, is unlikely to reach the Earth's surface for many millions more. The problem is that humans have found a shortcut to those natural geologic processes and have dug up the buried carbon and burned it to fuel civilizations. Shifting matter from one place in the Earth's system (underground) to another place in the system (the atmosphere) changes the nature of the entire system. As I explain in Chapter 6, systems balance the circulation of matter and energy to reach an equilibrium. Burning fossil fuels creates a positive feedback loop in the climate system, sending the system farther from equilibrium and causing other components of the system to shift in response.

The proof is in the isotopes

The most common question that people ask when confronted with all the scientific data about global climate warming is how scientists can be certain that humans are responsible for the increase in greenhouse gases, specifically carbon dioxide. The answer to this question is *isotopes*.

Defining isotopes and carbon ratios

An *isotope* is an atom of an element that naturally occurs with a different number of neutrons. For example, carbon most often has an atomic number of 12 (6 protons and 6 neutrons) and is labeled by scientists as *carbon-12* or ^{12}C. But other atoms of carbon exist with different numbers of neutrons (carbon-13 or ^{13}C has 6 protons and 7 neutrons and carbon-14 or ^{14}C has 6 protons and 8 neutrons). You may have heard of ^{14}C because it's the isotope that decays over time and that scientists use in radiocarbon dating.

Carbon atoms are constantly cycling through living and nonliving matter. When plants take carbon atoms from the atmosphere during photosynthesis (see Chapter 4), they absorb a greater number of ^{12}C atoms than ^{13}C atoms because the ^{13}C atoms are slightly heavier in mass and react a little differently in chemical reactions. Scientists measure the *isotope signature* of living and nonliving things by comparing ^{13}C to ^{12}C isotopes. This proportion, or *ratio*, of carbon isotopes ($^{13}C/^{12}C$) in plants is different from the ratio of carbon isotopes in the atmosphere. Because plants prefer the ^{12}C isotope, the ratio of $^{13}C/^{12}C$ in plants is lower than the ratio of $^{13}C/^{12}C$ in the atmosphere (where no biological process chooses one isotope over the other).

Burning plant material or fossilized plant material, such as fossil fuels, releases the ^{12}C isotopes that plants absorbed through photosynthesis millions of years ago into the atmosphere in the form of CO_2 molecules. The added ^{12}C isotopes in the atmosphere decrease the overall $^{13}C/^{12}C$ ratio.

Scientists compare the ratio of carbon isotopes in the atmosphere today to the ratio of carbon isotopes in the atmosphere over the last couple thousand years. (Fortunately, ice cores in Greenland and Antarctica have been trapping bubbles of atmospheric gas for thousands of years!) What scientists have found is that while the total amounts of CO_2 in the atmosphere have increased, the ratio of $^{13}C/^{12}C$ has gotten much lower since 1850 — right around the time when humans began burning fossil fuels for energy during the Industrial Revolution.

Shifting carbon ratios

Before humans began burning fossil fuels, the carbon ratios naturally shifted about 0.03 percent over a couple thousand years. Over the last 150 years (since humans began burning fossil fuels), the carbon isotope signature of the atmosphere has shifted more than 0.15 percent. Although this percentage appears small, it's five times more than the natural shift. Scientists consider a change that big to be *significant,* or beyond what would occur due to random chance.

A significant change means that something different is acting in the system to change things from what has been normal for the past 10,000 years or so. The something different in Earth's climate system is human industrial innovation and its ability to pull carbon from deep in the Earth and inject it into the atmosphere.

The question scientists are asking today is no longer *if* human actions have led to increasing greenhouse gases and a warmer climate but rather what now? What will a warmer climate mean for human civilization? And how can humans use their powerful influence on the Earth's climate system to conserve the resources and way of life they've depended on for thousands of years? To answer questions about how increasing GHGs will affect Earth's climate in the future, scientists look into the past to see whether a situation similar to today has ever existed before.

Considering Past, Present, and Future Climates

Earth hasn't always been the relatively cool place it is today with ice covering the poles. In fact, at times in its 4-billion-year history, Earth was much, much warmer. But for the last 2 million years or so, Earth has been cool enough to experience multiple ice ages.

When scientists look at Earth's climate history, they see that changes in the atmosphere, oceans, and continental ecosystems are closely linked. Fortunately, scientists can gather information about past climate conditions and use them to predict the future so that humans can better prepare for what's to come.

The ghost of climates past

Scientists called *paleoclimatologists* gather information about climates from long ago in many ways. Although they can't simply send a thermometer back in time, they can determine past temperatures on Earth by using *proxies,* or indirect measurements of temperature. Here are a few of the most common paleoclimate proxies that scientists use to determine past temperatures:

- ✔ **Ice cores:** Every year when it snows on the ice sheets of Greenland and Antarctica, a new layer of ice forms, trapping atmospheric particles of dust and bubbles of gas. Scientists use the layers of ice and the gas trapped in them to build an understanding of what the climate was like thousands of years ago. While they can directly measure the amount of carbon dioxide gas, they also use isotopes of oxygen as an indirect measure of temperature in the past.

- ✔ **Pollen:** Pollen grains produced by seed-bearing plants are made of special proteins that maintain their structure for millions of years if they're properly preserved — such as in the sediments of a lake bed. Scientists who study preserved pollen grains can identify the plants that produced the pollen. Knowing that certain plants thrive in certain conditions of temperature and moisture, these scientists can get a pretty good idea of what climates were like when these plants were alive.

- ✔ **Corals:** Some corals grow a new shell layer each year. As they build their shells from minerals dissolved in the seawater, they also record information about the water temperature and chemistry of the ocean they inhabit. Because ocean chemistry is closely related to atmospheric chemistry, scientists can use this information to reconstruct past climate conditions.

Measuring carbon dioxide levels in the atmosphere

Paleoclimate studies show that carbon dioxide levels in the atmosphere have varied throughout Earth's history. Scientists have a record of atmospheric CO_2 from ice cores that stretches back almost 500,000 years. When they compare the CO_2 levels in the atmosphere to temperatures across that same time span, they see that increases in CO_2 and warm temperatures go hand in hand, as Figure 19-3 illustrates.

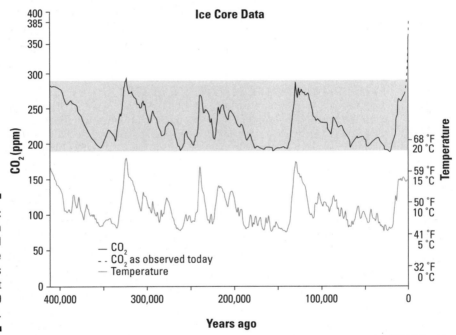

Figure 19-3: Carbon dioxide and temperature changes for the last 400,000 years.

Data: National Oceanic and Atmospheric Association (NOAA) Vostok Ice Core data and Mauna Loa CO_2 observations. Graph: John Streicker

Illustration by Wiley, Composition Services Graphics

This relationship between CO_2 levels and temperature makes sense with what scientists understand about how CO_2 (and other GHGs) affects atmospheric and surface temperatures.

Over the last 100 years, levels of CO_2 (specifically CO_2 molecules with isotope signatures that indicate they were created by burning fossil fuels) and other GHGs, like methane and nitrous oxide, have increased beyond the highest levels seen in the last 10,000 years. This increase means that GHGs and the Earth's climate are entering a phase unlike anything experienced by humans before.

Tracking patterns of variability

Studying past climates helps scientists recognize patterns or periodic shifts between warmer and cooler extremes. These patterns repeat themselves at different scales — some lasting thousands of years and others lasting less than a decade.

Patterns of change that repeat themselves in regular intervals are called *patterns of variability*. The term *variability* means that the changes aren't outside of what's considered normal but are simply part of a naturally occurring range of change. Keep in mind that patterns of climate variability are different from *climate change,* which refers to the overall climate trend, including the range of variability shifting to a new normal or average. In terms of modern climate change, this new normal will be a warmer range of temperatures than humans have experienced before.

Cycles of variability illustrate that climate isn't static or unchanging. But over long periods of time, sometimes thousands of years, boundaries for what's considered normal do exist. When the temperature and moisture conditions vary within the range of normal, scientists consider the changes part of natural variability. When the temperature and moisture conditions move beyond what has been defined over the last few thousand years as the range of normal, scientists begin to wonder what will come next.

The most common cycles of climate variability are Milankovitch cycles and the El Niño/Southern Oscillation.

Milankovitch cycles

Milankovitch cycles are patterns of variability in Earth's position relative to the sun. Because the Earth's orbit isn't always a perfect circle, sometimes the Earth is closer to the sun in its orbit than it is at other times. In addition, sometimes the Earth tilts more toward or away from the sun as it rotates on its axis. These changes occur cyclically over thousands of years and are linked to the recurring ice ages Earth has experienced in the last 2 million years.

El Niño/Southern Oscillation

The *El Niño/Southern Oscillation,* or ENSO, is a pattern of weather conditions that changes every few years in the Pacific Ocean. During years that are considered *El Niño years,* the Pacific Ocean water near the equator is warmer than usual. In years considered *La Niña years,* the Pacific Ocean surface waters are cooler than usual. As I explain in Chapter 7, the temperature of ocean waters is closely linked to ocean and atmospheric circulation patterns that circulate precipitation and heat around the globe. The cycles of El Niño and La Niña are associated with specific changes in local weather patterns in regions surrounding the Pacific Ocean, such as the west coasts of South America and North America.

Predicting the future: Climate models

Predicting the climate future isn't as simple as looking into the past, but scientists can use what they've learned about how GHGs, atmospheric circulation, and ocean circulation affect global climate to create models, called *climate models,* that predict future climate conditions. (As I explain in Chapter 2, *models* are useful tools that scientists use to help them better understand complex interactions and test hypotheses.)

So how do climate models work? They're basically just computer programs that explore how a collection of climate variables interact with each other. The interactions in each model are based on real interactions observed by scientists, but in the model, scientists can change the variables to test different scenarios of the future.

Some models are very specific, focusing only on atmospheric circulation patterns, for example. But the most powerful models are *general circulation models* or GCMs, which combine models of atmospheric circulation and ocean circulation to get a picture of how heat and moisture are circulated throughout Earth's entire surface environmental system.

When scientists construct a climate model, the first thing they do is test it by using variables from a known scenario — for example, using data from the last 20 years to predict this year's climate conditions. Scientists work to adjust the model interactions until they simulate reality as much as possible. When they trust that the model is as real as possible, they can use it to predict future outcomes with trustworthy results.

Most climate models predict that the high levels of GHGs in the atmosphere are going to increase temperatures on Earth from 4 to 6 degrees Celsius over the next 100 years. Although 4 to 6 degrees may not seem like much — you probably experience that kind of temperature change in one day during some seasons — an increase of 4 to 6 degrees on average in the entire global system will undoubtedly have dramatic effects on the planet and every living thing that inhabits it (including you).

Feeling the Heat: Environmental Effects of Modern Climate Change

The changes in the Earth's system that scientists have already observed and attributed to climate warming include changes to the atmospheric chemistry, ocean chemistry, weather patterns, and the distribution of life across the continents. The effects of a changing climate are already evident in ecosystems across the globe, and as temperatures get warmer, modern climate change will increasingly affect human societies around the world.

Here are just some of the observed and expected effects of global climate change:

- **Rising sea levels:** As you know, ice needs to be cold or it'll melt. In a world where temperatures are gradually rising, that simple fact means that ice, where most of Earth's fresh water is currently stored, is going to melt. Warmer temperatures are already affecting the amount of ice at the North and South Poles. Over the last 30 years, the amount of sea ice (floating ice on the ocean) present across the Arctic each winter has shrunk. Shelves of ice in the Antarctic are shrinking as well. Melting ice at the poles adds fresh water to the oceans, raising the sea level relative to land anywhere from a few inches to a few feet, depending on the location and extent of melting. Even a slight rise in sea levels causes major problems, such as saltwater intrusion to freshwater wells and inundation of low-lying coastlines around the globe (see Chapter 16).

- **Melting permafrost:** In polar regions, much of the ground is saturated with water and frozen all year long. This *permafrost,* as it's called, is melting and will continue to melt as global temperatures continue to rise. Melting permafrost increases the amount of both carbon dioxide and methane in the atmosphere as organic material that was frozen decomposes. Adding these GHGs to the atmosphere creates a positive feedback loop in the climate: Warmer temperatures lead to more melting, which leads to more methane and carbon dioxide being released, and this increase in GHGs creates a more intense greenhouse effect, which creates even warmer temperatures, and so on and so on. (See Chapter 6 for more on positive feedback loops.)

- **Shifting weather patterns:** Climate models suggest that the variability in weather patterns will become more extreme, in particular, that warmer regions will experience more heat waves. Precipitation is the most difficult weather aspect for climate scientists to pin down, but many models suggest that rainfall will increase dramatically in already-wet regions — enough to cause flooding and landslides — while currently dry regions will become even drier and suffer increased drought and water stress.

✔ **Increasingly acidic oceans:** Ocean pH varies around the world, but seawater has, on average, a high pH value, making it slightly basic (see Chapter 3 for details on pH). Carbon dioxide is constantly moving between the atmosphere and the ocean. Increasing levels of CO_2 gas in the atmosphere send more CO_2 into ocean waters. When CO_2 dissolves in ocean water, it changes the pH of that water, making it more acidic. Even small changes in the pH of ocean water are enough to affect the marine ecosystem. In particular, more acidic ocean water contains fewer dissolved carbonate compounds available for sea creatures to use when building their shells, which leaves them vulnerable to predators and generally less protected. While some organisms may adapt to a changing pH, others may not be able to adapt quickly enough to keep up with the speed of acidification scientists are currently recording. Figure 19-4 illustrates the complex relationship between carbon dioxide in the atmosphere and the ocean.

Changes in the global climate affect ecosystems and the organisms living in them in specific ways. For example, because of modern climate change, the *biomes* (or broadly defined ecosystem categories) that organisms are best adapted to are likely to shift in location or disappear altogether (as may be the case in the Arctic). As a result, the organisms that live there will be left to quickly adapt or face extinction. (See Chapter 7 for more on biomes.)

Scientists observe one example of ecosystem disruption due to warmer temperatures in the Netherlands with a species of bird called the *pied flycatcher*. The pied flycatcher chicks hatch each year at the same time of year that caterpillars are most abundant. The chicks feed on the caterpillars for the first few days of life when they need a large amount of nutrition. In recent decades, however, warmer temperatures have encouraged tree leaves to grow earlier in the season. Because tree leaves are the caterpillars' main food source, the caterpillars have started to come out sooner so they have plenty of food to eat. By the time the pied flycatcher chicks hatch a few weeks later (because they're unaffected by the warmer temperature), no caterpillars are left for them to feed on.

Coral reef communities have also seen ecosystem disruptions in response to warmer water temperatures. One example is *coral bleaching,* which occurs when environmental stress disturbs the symbiotic relationship between the coral animal and the photosynthetic microorganisms that live inside it (see Chapter 8 for details on symbiosis). A shift in environmental conditions (in this case, temperature change) interferes with the coral's ability to support the microorganisms, and the symbiotic relationship ends. Because both organisms depend on their relationship to survive, both species suffer population decline. These declines have dramatic effects on the reef ecosystem, which depends on corals for ecosystem structure.

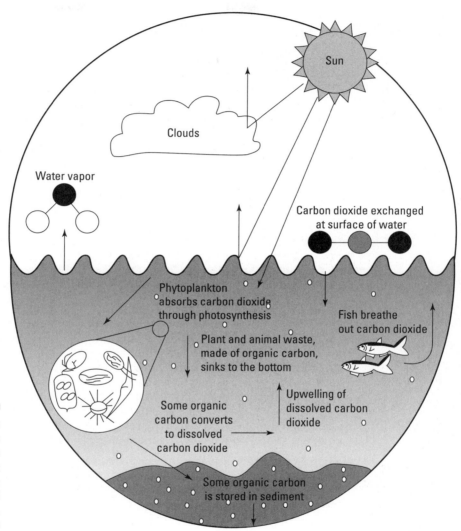

Figure 19-4:
The relationship between carbon dioxide and the oceans.

Illustration by Wiley, Composition Services Graphics

The pied flycatcher and coral bleaching are only a few examples of how intricately different species in an ecosystem are linked to one another and how some species are effected by temperature changes. As the world's climate continues to warm, scientists will observe more changes like these in the world's ecosystems — changes that will either require adaptation or lead to species extinction.

Facing a Warmer Future

Human beings aren't immune to the ecosystem disruption and other challenges presented by a warming climate. Unlike other organisms, however, humans do have the ability to understand part of what's happening, predict how it will affect human society in the future, and take steps to minimize the damage or adapt. In this section, I describe how modern climate change is likely to affect humans and what scientists and policymakers suggest doing to reduce its impacts and prepare for a warmer future.

Feeling the effects on human society

As a result of global warming, human beings will face challenges of adaptation. Fortunately for some human societies, technological innovation will protect them from changes in their immediate environment. For example, if you already have air conditioning and next year the summer heat is a little hotter than usual, you can simply turn up the air conditioning and wait out the heat.

However, billions of human beings still live closely in touch with their immediate environment and will be affected by even small shifts in temperature and precipitation. Some of the projected effects on human society include

- **Drought and water stress:** Many regions of the world already struggle to maintain enough freshwater supplies for the existing human population. If patterns of precipitation shift in a way that leaves these regions even drier than they were during the last 100 years, major droughts will occur, devastating farmers and those who rely on their crops for food and causing people to relocate in search of fresh, clean water.

- **Changes in food security:** Shifting ecosystems mean that where farmers currently grow crops may no longer be suitable for those crops 50 years from now. On the one hand, this means that new regions will open up to agriculture, but on the other hand, it means that the stability of worldwide food resources in the next 50 years is unknown.

- **Increasing cases of certain diseases and illnesses:** In particular, scientists are concerned about the spread of diseases by ticks, mosquitoes, and other insects as these insect species follow warming temperatures and move into new regions of the world.

- **Shifting tourism economies:** Parts of the world are known for their tourist attractions; over the last century, people who live in these regions have built entire cities supported by seasonal attractions such as winter sports or seaside vacations. Warming temperatures will reduce the presence of snow in some regions, and rising sea levels will change the nature of coastlines and beaches. In both cases, these locations will need to adapt to a new economic base if they want to survive.

✔ **Increasing conflict over available resources:** With water and food stress, particularly in countries without stable governments that can provide for those in need, comes conflict. As different groups of people struggle to provide for the basic needs of their families, many will resort to violence. This issue has already been seen in parts of Africa that have been experiencing drought for the last few years.

Turning down the heat: Mitigating human influences

Now that more and more people understand that human actions influence the climate system and cause global temperatures to get warmer, individuals, communities, and entire nations are taking steps to reduce, or *mitigate,* human impacts.

One important mitigation method involves reducing carbon dioxide emissions and finding ways to store carbon somewhere other than the atmosphere in places called *carbon sinks.* Carbon sinks include soils that are made up of plant biomass in forests and in the deep ocean. Taking carbon out of the atmosphere and putting it into a carbon sink is called *carbon sequestration;* the climate system can do this naturally or humans can do it artificially. Artificial sequestration methods include using carbon scrubbers, or filters, to remove carbon dioxide from the air and then pumping it into underground storage spaces or planting forests to increase carbon storage as plant biomass.

In the 1990s, multiple nations recognized the global threat of a warming climate. In 1997 at a meeting in Kyoto, Japan, these nations agreed to a plan for reducing GHG emissions in developed nations (leaving China and India the room for continued economic development through coal burning). Their goal was to reduce emissions by 2012. The agreement, called the *Kyoto Protocol,* was signed by 160 countries (not including the U.S.).

The largest contributors to carbon dioxide in the atmosphere are the industries and power suppliers that depend on coal burning for energy. Encouraging industries and nations that produce large amounts of atmospheric carbon dioxide (and other GHGs) to reduce emissions has the potential to make a difference in the warming climate trends that scientists observe and predict. With this in mind, nations have held multiple conferences since Kyoto to address climate change as an international challenge. The meetings have met with varying levels of cooperation and agreement among nations.

The U.S., for example, didn't agree to the Kyoto Protocol, but individual cities and states within the U.S. have taken it upon themselves, without federal guidelines or assistance, to meet Kyoto Protocol standards and carbon efficiency.

Another approach to reducing human impacts on the climate is to encourage each person to take responsibility for his or her personal carbon output. Start by quantifying your carbon footprint. Similar to an ecofootprint (see Chapter 5), a *carbon footprint* calculates how much carbon you produce in your daily life. The calculation includes things such as your method of transportation and daily energy use as well as the fossil fuels used to transport food and other goods to stores where you purchase them. After you're informed about how your choices result in carbon dioxide emissions, you can determine where you can most effectively or reasonably reduce your personal carbon footprint. Check out the website www.carbonfootprint.com to calculate your carbon footprint and find ideas for reducing your carbon impact.

Engineering solutions to global warming

As politicians continue to debate whether to take legislative action toward reducing human impacts on the global climate system, some scientists are calling for drastic measures and creative, *manmade* solutions to this *manmade* problem.

Using technology to change climate conditions on Earth is called *geoengineering.* Geoengineering solutions usually tackle climate change one of two ways: by capturing carbon or by controlling solar radiation input. Ideas range from simple to imaginative/extravagant and include the following:

✔ Painting roofs and other surfaces light colors (like white) to increase *albedo* (reflection of the sun's energy)

✔ Capturing carbon from the air with special filters and burying it back in the Earth

✔ Building a solar sunshade, or reflective mirrors, outside of Earth's atmosphere to decrease incoming solar radiation

✔ Adding iron to the oceans so they can absorb more carbon and store it as sediment on the seafloor

✔ Creating a manmade "volcanic eruption" that fills the upper atmosphere with enough sulfur particles to create a solar-blocking haze

✔ Installing pipes in the ocean to help circulate cooler water up from the ocean depths

Some people like the idea of using technology to solve the global warming problem, while others think the focus on elaborate geoengineering plans for the future detracts from working to reduce emissions now. The simplest geoengineering ideas (such as painting rooftops white) certainly can't hurt. But it's still too early to tell whether the more elaborate plans will be safe and effective or have negative consequences. After all, scientists can't be sure what effects the more imaginative plans will have until they've tested them on the Earth, its climate, and its ecosystems.

Adapting along the way

Reducing human impacts on climate change is important, but scientists recognize that a shift in the Earth's overall climate system is already in motion. This shift doesn't mean a catastrophe is imminent, but it does mean that although humans are now taking steps to slow down humanity's disproportionate influence, they must also begin adapting to a climate that modern human societies have never experienced before.

In terms of climate change and humans, *adaptation* means conserving resources or finding new resources to meet the needs of human beings. Some ways humans may adapt to climate change in the coming decades include

- ✔ Increasing energy efficiency and diversifying energy resources to shift from a dependence on fossil fuels

- ✔ Increasing water efficiency and conservation, especially in drought-prone or water-stressed regions

- ✔ Building infrastructure to protect coastlines and coastal urban centers from sea level rise

- ✔ Adjusting crop rotations or crop species to coincide with shifting seasonal conditions

- ✔ Creating protected corridors for wildlife migrations in response to shifting biomes

- ✔ Boosting public health resources and education to prepare for increased spread of certain diseases

Part V
Follow the Recycled Brick Road: A Sustainable Future

The 5th Wave   By Rich Tennant

We're reducing GHG emissions on the ranch by switching to all electric cows.

ELECTRIC CHARGING STATION

In this part . . .

All environmental scientists have their eyes on the future, thinking long term about conserving environmental resources so that they'll last. Living more sustainably requires a shift in how you think about your place in the environment and how you value the resources, products, and services the environment provides.

In this part, you find out some basic economic ideas about how to include environment health in calculating the cost of everyday products. This part also describes the most important and effective laws for protecting the environment, including international agreements. After all, there's only one Earth.

Chapter 20

Building a Sustainable Future

*Y*ou may hear the argument that sustainability just isn't economically sound, but the truth is that many of the current methods of extracting resources and manufacturing goods aren't sound. Mainstream economics relies on the assumption that Earth's natural resources are limitless, and that premise simply isn't true. Some resources, such as fossil fuels, are limited in supply; other resources, such as fresh water, become more limited as they're overused or polluted. Because of resources' limited nature, economic growth based on an idea of unlimited supply is unsustainable.

By practicing sustainable methods of resource extraction, production, and development, humans can meet their endless desire for resources with the limited supplies the Earth has to offer. However, achieving this level of sustainable living requires a shift in both how humans view their role in the global ecosystem and how economists measure wealth.

In this chapter, I go over some common worldviews and explain how shifting your perspective can help you understand the effect your actions and choices have on building a sustainable future. I describe a few basic principles of economics and discuss how wealth is measured. I explain how adding environmental cost into economic equations can help people and businesses make more sustainable choices and how some economists have begun thinking of the economy like an ecosystem. Finally, I cover the close relationship between poverty and environmental degradation, including steps that can reduce poverty and improve the environmental health for many of those living with only a bare minimum of life's necessities.

Overlapping Spheres of Sustainability

The word *sustainable* appears throughout this book because sustainability is arguably the most important concept that human culture has stumbled upon for the last few thousand years. In terms of environmental science, *sustainability* is the idea that humans can use and manage natural resources so that those resources can provide for human needs for as long as possible (potentially forever).

Sustainability is often described as having three *spheres,* or areas of influence. Figure 20-1 illustrates these spheres of sustainability.

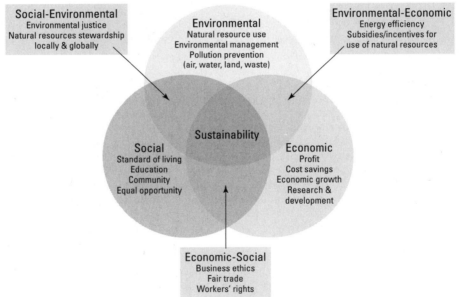

Figure 20-1:
The three spheres of sustainability.

Adapted from 2002 University of Michigan Sustainability Assessment
Ilustration by Wiley, Composition Services Graphics

- ✔ **Social:** The *social sphere* of sustainability includes educating, community building, and providing equal opportunity to everyone to live in a clean, healthy environment.

- ✔ **Economic:** The *economic sphere* of sustainability includes the economics of sustainable growth, the production and consumption of goods and services, and the funding of research into sustainable methods of production and technology.

- ✔ **Environmental:** The *environmental sphere* of sustainability includes natural resource management, environmental protection, and conservation.

The most important parts of Figure 20-1 are where these three spheres overlap. For example, where the social and economic spheres overlap, you find the issues of workers' rights and fair-trade production. The overlap between economic and environmental is where regulations and economic incentives for environmental protection come in. Where the social and environmental spheres overlap is where you find the concepts of social justice (which I describe later in this chapter) and *stewardship* (responsible management of natural resources).

The goal for many professionals working in environmental science is to find ways for these three spheres to overlap cooperatively toward the common good.

Looking at Ecosystem Lessons

Environmental scientists and economists who study sustainable development point out how much people can learn about sustainability by simply observing natural ecosystems. Three features of ecosystems in particular stand out as lessons to humans about sustainability; sustainable societies incorporate these ecosystem lessons into their thinking about economic development. In an ecosystem

- **One creature's waste is another creature's food.** A natural ecosystem doesn't produce trash that must be stored and that serves no other purpose. A sustainable human society finds a way to use its waste as fuel or to create new products from that waste instead of just finding more efficient ways to store it as trash.

- **Energy is supplied by the sun.** The sun's energy is the closest thing Earth has to an endless energy supply. Almost every plant and animal on Earth's surface is fueled by energy captured from the sun through photosynthesis (see Chapter 4 for more on how photosynthesis works). A sustainable human society finds ways to harness the sun's energy as solar, water, or wind power. (Chapter 14 discusses ways to harness the sun's energy and other alternative energy resources.)

- **Diversity provides stability.** Ecosystems with a wide variety of life forms are more resilient to disturbance and provide multiple options for adaptation in the face of change. Respecting the variation of individuals, species, and ecosystems and working to preserve that diversity provides humans with a stronger, more stable base of natural resources to meet their needs.

Counting Coins: Mainstream Economics

Economics is the study of how goods and services are produced, distributed, and consumed. Obviously, this book isn't about economics, but understanding some economic terms, concepts, and principles is useful to environmental scientists because the use (and sometimes destruction) of natural resources is a result of the human need or desire for goods and services.

In this section, I define some basic economic terms and contrast how mainstream and environmental economists calculate wealth.

Measuring Earth's capital

In economics, wealth is created by *capital,* a product or item that builds more wealth. Raw materials such as timber and metals are capital, as is money, which can be invested in the production of goods to increase capital. Many different types of capital are available, including the following:

- **Built capital:** *Built capital* includes roads, buildings, technology, and tools that are used to produce goods and services and increase wealth.

- **Human capital:** *Human capital* is knowledge, valuable experience, or education used to produce goods and services and increase wealth.

- **Natural capital:** *Natural capital* includes the goods and services nature provides that are used to increase wealth. For example, a river provides water for crops, supporting agricultural production.

Environmental economists think that a true calculation of wealth requires more accurately estimating or quantifying natural capital. Considering the true value of natural capital should lead producers to desire more sustainable methods of resource extraction and production — ways to invest in natural capital as well as human and built capital.

Going to the market

Economically speaking, a *market* is the trading of goods, services, or money. A market relies on supply and demand; *supply* refers to the products offered for trade, and *demand* indicates the number and type of product that consumers want to acquire. Supply and demand reach an equilibrium when the price is right — high enough that providing exactly the amount of goods or services that the consumer demands is financially worthwhile for the seller.

When you purchase something at the store, you pay what's called the *market price*. Currently, the market prices of most goods and services don't include a calculation of the products' environmental effects. For example, when you buy paper, the price you pay covers the cost of raw materials, labor, manufacturing, packaging, and maybe the fuel to deliver the product to your local paper store. What it doesn't include is the cost of air pollution created during the manufacturing process or of the soil erosion, water pollution, and habitat destruction that deforestation causes.

These environmental costs are considered *external costs* or *externalities*. They're not included in the market price because they're difficult to quantify, and producers (especially those that cause high levels of environmental harm) fear that doing so will increase prices and possibly reduce consumer demand. Currently the planet's environment carries the burden of these costs, while the product manufacturers benefit.

Calculating a nation's wealth

Currently economists measure the wealth of a nation or economy by the *gross domestic product,* or GDP, which is a calculated value for all the products the nation produces. GDP calculations don't include the cost of externalities (see the preceding section) and, therefore, don't truly reflect the production cost for a nation's products. Traditionally, economists have also considered a nation's overall consumption to be an indicator of wealth — the more resources a nation consumes, the wealthier the nation is.

Some scientists have suggested that a more accurate measure of a nation's economic well-being is the *genuine progress indicator,* or GPI, which measures characteristics of a nation beyond its products, including people's health, levels of education, and consumption levels, along with the amount of pollution and environmental degradation they experience. When measured from this perspective, many nations considered wealthy by GDP standards are experiencing a declining well-being overall.

Tripling the Bottom Line

Environmental scientists and environmental economists suggest that externalities be included in the calculation of a product's market price. Adding the cost of social and environmental effects into the market price of an item is called *tripling the bottom line.* This cost reflects the real value of the item in terms of resources, environmental quality, and human health.

You may already see the difference in price for goods that have included some external costs in their market price — for example, the price difference between fair-trade coffee and regular coffee. The regular coffee is much cheaper, but that's because the fair-trade coffee factors into its market price the cost of using sustainable practices to keep from damaging the ecosystem and to be sure that farmers and farmworkers harvesting the coffee are paid a fair living wage.

Figuring out how to add the cost of pollution and environmental damage to the price of a product is tricky. In the current U.S. economic system, it requires laws or regulations and a regulatory agency to make sure companies follow the rules. The following sections introduce a couple of different ways that economists propose to include environmental costs in the market price of a product or service.

Taxing harmful practices

Some environmental economists believe that taxing environmentally damaging practices is a surefire way to encourage companies to rethink those activities. When a company is faced with such a tax, it may find that the long-term benefits of shifting to more sustainable practices far outweigh the initial cost of such changes. An *eco-tax* or *green tax approach* means that instead of taxing labor and profits, authorities tax the specific activities that result in pollution and environmental damage; one example is quantifying the amount of air pollution emitted from the manufacture of a product and placing a tax on those emissions. The company either pays that tax or passes the fee on to the consumer, which means the product will cost more.

Instituting incentive-based regulations

Incentive-based regulations encourage producers to use sustainable practices and reduce waste in production.

An example of an incentive-based system is the *cap and trade system* for air pollution. The U.S. Environmental Protection Agency (EPA) instituted a cap and trade regulation system to tackle the problem of acid rain (see Chapter 15 for details on acid rain and air pollution). The EPA Acid Rain Program does the following:

  ✓ Sets a maximum limit (the cap) on the amount of acid rain-causing air pollutants (sulfur dioxide and nitrous oxides) that a polluter can release into the atmosphere.

✔ Requires tracking and reporting of the air pollution produced.

✔ Provides a market where polluters can trade credits if they haven't used their maximum allowance of air pollution. For example, a low-emission producer can sell its leftover share of allowed pollution to a high-emission producer.

The goal of the EPA Acid Rain Program is to encourage coal-burning plants to significantly reduce air pollution. Similar cap and trade plans have been suggested to combat carbon dioxide emissions (and global climate change) in the worldwide market.

This type of tradable pollution credit has advantages and disadvantages. Although the system is flexible and encourages companies to reduce pollution, it also provides a way for the largest polluters and wealthiest companies to work around environmental regulations without changing to more sustainable practices.

Exercising the Precautionary Principle

The drive to create more wealth through consumption ignores the potential risks of some resource extraction, production, or waste management methods. Environmental scientists suggest that industries practice the *precautionary principle,* which states that if a product is potentially harmful to the environment or to human health, the risk of harm isn't worth it; a better choice is to go with an option known to be harmless.

This guideline sounds reasonable enough, but it means that every product must be proven harmless before it can go on the market for consumers. In contrast, U.S. industries currently practice the *philosophy of innocent until proven guilty,* which means that a product is removed from the market only after it has been proven to cause harm. This approach allows companies to put more products on the market more quickly, but it also means that they don't have to pull a product from the market until some harm has not only already occurred but also been scientifically proven.

Some countries already practice the precautionary principle, removing potentially harmful products from the market until scientific tests can be completed to determine whether the products are dangerous. For example, in 2010, Canada listed bisphenol-A (BPA), a common compound in plastic bottles and plastic-lined cans, as a toxin. This move made selling products that included BPA illegal. Research continues on exactly whether and how BPA affects humans, with some early studies suggesting it functions as an endocrine disruptor (see Chapter 17 for details on this kind of toxin). The

Canadian government chose to exercise the precautionary principle and remove BPA from the market while scientists continue to research its effects. In contrast, U.S. regulators have chosen to wait until further scientific study proves BPA is harmful before creating strict regulations guiding its use in consumer products. As a result, if BPA is proven to be a harmful toxin, billions of people will already have been exposed to it for decades, with potentially irreversible effects.

Looking at the World in New Ways

Three main viewpoints describe how people perceive the world and the role of human beings in that world:

- **Anthropocentric:** An *anthropocentric* worldview places human needs and desires as the top priority. People who approach the world from this perspective accept that natural resources, including other living creatures, are meant to be used to meet human needs.

- **Biocentric:** A *biocentric* perspective considers human beings as one of many different species on Earth, each of which has a right to exist. The role of the ecosystem in this worldview is to support living creatures.

- **Ecocentric:** An *ecocentric* worldview equally values living things and the ecosystems they inhabit. From this perspective, humans are one of many creatures living within and depending on ecosystems and their resources.

You may find that you don't strictly adhere to any of these perspectives or that your perspective differs depending on the issue. For example, perhaps you're more anthropocentric in your need for energy resources but think more biocentrically about conservation of biodiversity.

Shifting from mainstream economic ideas about how to grow and measure wealth to sustainable ideas about how to best use natural resources and calculate wealth requires people to look at the world from a new perspective. Some economists have begun to view the world more ecocentrically. In particular, they seek to understand the global economy as a system that can provide for human needs, similar in many ways to an ecosystem. I describe these shifts to more sustainable ideas in the following sections.

Evolving economics

Economic analysis that includes the cost of environmental damage and degradation of natural resources is called *environmental economics*. Environmental economics seeks to include the cost of environmental damage in product pricing through taxes, fines, or regulations such as the ones I describe in the earlier section "Tripling the Bottom Line."

Ecological economists take another approach. *Ecological economics* looks at how ecosystems function in terms of systems, energy, and matter cycling (see Chapters 3, 4, and 6) and includes this understanding in the study of economic systems. Ecological economists recognize that the natural world is part of the economy and that it provides raw materials for goods as well as ecosystem services and natural capital.

Conventional economic analysis, on the other hand, considers many ecosystem services and natural capital free for the taking. For example, people don't pay for a wetland that filters, cleans, and restores groundwater supplies. But when the wetland disappears or the groundwater becomes contaminated, those people face costs in terms of human health and the need to replace that ecosystem service.

Ecological economists use a process called *valuation* to give ecosystem services and natural capital a cost or price in terms of money. Sometimes valuing the environment can be straightforward — for example equating the value of a wetland that filters and cleans groundwater to the technology needed to complete the same purification process. Other times, though, it's much less direct; what's the value of each polar bear?

Valuation of natural resources and ecosystem services is one of the biggest hurdles to shifting how people think about sustainability. Sustainable choices in products, services, and business practices are often painted as economically unrealistic for continuing growth and increasing wealth.

Placing a value on ecosystem services and natural resources provides a more complete basis for cost-benefit analysis on the goods and services you purchase. *Cost-benefit analysis* is an economic tool that compares the cost of something to its outcome or benefit; it's used in decision making to determine what's worth investing money and resources in based on whether the benefits outweigh the costs. Including the cost of environmental damage in the cost-benefit analysis may dramatically change a product's price.

Reevaluating production and consumption

Creating a sustainable economy requires that producers and consumers rethink their roles in the *production-consumption cycle*. Figure 20-2 illustrates this cycle, as well as all the places within the cycle that pollutants and waste are created.

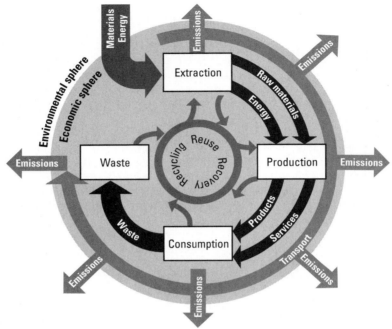

Figure 20-2:
The
production-
consumption
cycle.

Illustration by Wiley, Composition Services Graphics

The production-consumption cycle begins with extracting raw materials. These materials are transformed into products sold to consumers. After you, the consumer, are done using the product, the product is considered waste. This cycle of production and consumption moves in one direction, is considered a *linear production-consumption cycle,* and is what produces the waste stream that I describe in Chapter 18. The result of this linear approach is that raw materials and natural resources are transformed into trash that must be stored somewhere and, in many cases, contaminates and pollutes the environment. (Flip to Chapter 18 for more on waste and its disposal.)

An alternative to this wasteful approach is a *closed-loop production cycle,* or *closed-loop recycling.* In this system, the consumer products are recycled and used to create a new or different type of product instead of being stored as garbage. Ideally, the original raw materials are transformed into useable products repeatedly, which reduces the need to gather new raw materials and reduces the amount of waste that must be disposed of in landfills.

Figure 20-3 illustrates a closed-loop production cycle.

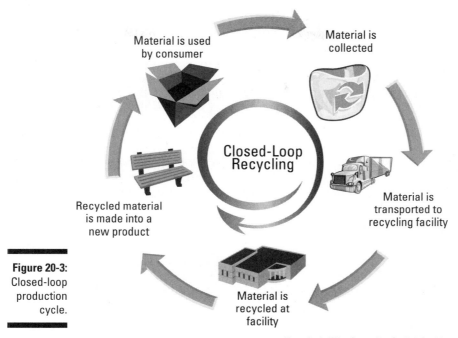

Material is used
by consumer

Material is
collected

Closed-Loop
Recycling

Recycled material
is made into a
new product

Material is
transported to
recycling facility

Figure 20-3:
Closed-loop
production
cycle.

Material is
recycled at
facility

Illustration by Wiley, Composition Services Graphics

It Takes a Global Village: Poverty and Sustainability

As the human population grows and nations continue to seek increased wealth, considering how Earth's resources are cared for and conserved for future generations becomes even more important. Preindustrial nations, where many of the poorest people live, can learn from the mistakes of more developed nations. Lifting people out of poverty through sustainable development is a powerful way to create a brighter future for everyone.

In this section, I explain how poverty and the environment are closely linked and describe some of the ways reducing poverty and increasing sustainable development all over the world result in healthier ecosystems and wealthier societies.

Seeking justice

Many environmental scientists believe in social justice. *Social justice* is the idea that the world has enough resources to meet the needs of everyone but that financial inequality and unequal resource distribution result in hunger, poverty, violence, and environmental damage. Social justice advocates that the Earth's resources be managed in a way that provides for everyone's basic needs.

Almost 2 billion people live in poverty around the world. People living in extreme poverty don't have the option of choosing the more sustainable path; they're struggling each day just to get enough clean water and food. (When economists speak of *poverty,* they're referring to people who are unable to meet their basic needs for food, fresh water, and clean living conditions.) Often, these poorest of the poor are the ones who degrade natural resources through overuse — farming until the soil is completely emptied of nutrients, grazing cattle until no grassland remains, and so on.

Environmental scientists and social scientists realize that improving conditions for people living in poverty improves not only those people's lives but also their environment. Examples of such simple ways to improve the lives of those living in poverty include the following:

✔ Providing basic education for children

✔ Battling malnutrition and disease

✔ Educating women about family planning to reduce birth rates

✔ Investing in local-scale sustainable infrastructure such as solar-powered energy grids and sustainable agriculture

As I explain in Chapter 8, increasing human populations, just like populations of other organisms, can put a strain on their environment. More people mean the need for more resources. Educating women and improving health conditions leads to slower population growth. Couple these advances with investing in sustainable infrastructure, and you pave the way for a sustainable and much improved way of life for those struggling to meet basic daily needs.

Social justice is related to environmental justice. *Environmental justice* means that all people have a right to be equally involved in making decisions about their environment. Advocates of environmental justice work to ensure that poor people and minorities are not disproportionately burdened with pollutants and other environmental damage.

Microlending a hand

One way to reduce poverty and encourage social justice is through *microlending*. The idea of microlending is simple: Rather than depend on large investments in national infrastructures to change living conditions for poor people, microlending offers small loans that can be used to make small changes with large impacts. For example, a woman who sells eggs in a village market can use a microloan to purchase more hens or build a larger henhouse to protect her chickens. The amount of money required for this small improvement is tiny, but it can have dramatic results, bringing more wealth to this woman and her family.

Microlending got started in the 1980s and is still a relatively new concept for many people. Microloans are risky because you're loaning money to people who don't have the wealth to ensure they'll pay you back anytime soon. One reason that microloans are successful is that average people in wealthier, industrialized nations can afford to offer microloans; a minimum loan amount may be $25. Organizations such as www.kiva.org help lenders locate borrowers in need around the world and coordinate loan disbursement and repayment.

Chapter 21

Cooperating to Sustain the Earth

Considering that millions of people in almost 200 different nations have to share the Earth, finding a way to sustainably share and protect its resources can be a little tricky. One way to protect and manage environmental resources is through government regulation. However, the world doesn't have a universal governing body that can pass laws that apply to every nation. Instead, groups of nations come together to create agreements, such as treaties, to regulate the use and protection of global resources. Some of these international agreements have made a major difference in the health and well-being of ecosystems and resources around the world.

In this chapter, I briefly describe the major players in environmental regulation both internationally and within the U.S. I also explain some of the most important laws and agreements that have been created to govern Earth's shared resources and environment. These regulations have met with varying success, but some of them serve as proof that the international community can unite toward the goal of global sustainability.

Understanding What's at Stake

The only way to successfully manage and share the water, air, and other resources on Earth is for people everywhere to work together toward this common goal. When multiple people or groups of people share an interest in a resource of some kind, they're called *stakeholders*. Each stakeholder cares about the resources, but they may have very different reasons for caring. For example, when the issue of timber resources (in other words, forests) comes up, many different stakeholders want to be part of the decision-making process. These stakeholders may include the timber industry, the paper industry, water conservationists, forest ecologists, hikers, and wilderness advocates, just to name a few. Each of these stakeholders has a different perspective on why forests are important, but they have one thing in common: an interest in keeping the forests healthy and productive.

In talking about global natural resources, such as the oceans and the atmosphere, every nation on Earth becomes a stakeholder, and each one has a different idea about which solutions best suit its needs. Each nation must consider many things in the decision-making process, including the health and well-being of its citizens, the potential for economic growth, and the need for raw materials and energy.

Stakeholders with very different priorities often have a hard time creating agreements that make everyone happy. In the next section, I describe a few of the agreements that the international community has created concerning global environmental resources and *stewardship,* or caring for the Earth.

Uniting Nations: International Agreements on Planetary Stewardship

Although no governing body regulates the actions of every nation on Earth, different nations have come together on their own accord to create international agreements about how to maintain and care for the Earth's natural resources. These international agreements and treaties are often drafted during large meetings, or conventions, that representatives from various interested nations attend. In this section, I describe a few of these international conventions, the resulting agreements, and their outcomes.

The Ramsar Convention on Wetlands

The goal of the *Ramsar Convention* is to protect and conserve wetland resources. In 1971, representatives from multiple nations met in the Iranian city of Ramsar and signed a treaty that recognized "wetlands of international importance." In this case, the term *wetland* includes all water habitats and biomes except the oceans. Using an ecosystem-centered approach and implementing sustainable use practices, the Ramsar Convention describes how each nation and the international community together can protect wetland habitat all over the world by following these recommendations:

✔ Create national wetland policies within each nation.

✔ Consider the traditional and cultural value of wetlands.

✔ Encourage sustainable use practices to support water quality, fisheries, wildlife habitat, agriculture, and recreation.

✔ Strengthen community involvement in wetland conservation.

✔ Increase education and knowledge of wetland resources.

The convention members continue to meet regularly in countries around the world to revise and expand wetland protection in the international community and update the Ramsar list of important global wetlands. Check out www.ramsar.org for details on the convention's current activities and recent meetings.

The Convention on International Trade in Endangered Species

The *Convention on International Trade in Endangered Species,* or CITES, is one of the largest international agreements concerning the environment. Since its creation in the 1960s, CITES has succeeded in preserving endangered or threatened species all over the world. Members of this agreement work together to govern the trade in endangered species and to make sure that animals and plants in danger of becoming extinct aren't harmed by international trade that further diminishes their populations. One of their most difficult, ongoing struggles is to keep poachers from illegally harvesting animal products, such as elephant ivory and rhinoceros horns. Check out www.cites.org for details on their work concerning endangered species around the world.

The Convention on Biological Diversity

The *Convection on Biological Diversity* (CBD) is an international convention whose focus is on preserving global biodiversity in the age of genetically modified organisms. The *Cartagena Protocol* on Biosafety is an agreement created by members of the CBD to safely control the transportation of genetically modified organisms. The protocol describes guidelines for transporting living things that have been genetically modified or are otherwise the results of biotechnology to avoid potential ecosystem disruption or endangering human health.

The Cartagena Protocol has achieved mixed results. While it has been successful at creating a forum for international discussion of biosafety issues, members are still developing ways to implement and enforce its guidelines. Go to www.cbd.int for information on the convention and news articles related to issues of biosafety.

United Nations agreements

The *United Nations* (UN) is an international organization with members from nearly every nation in the world. Formed after World War II with the goal of providing a place for discussion and resolution of conflict without war, the UN has also played a large role in other international agreements. In fact, some of the most important international agreements concerning the environment are the result of UN conventions.

Stockholm Declaration

The *Stockholm Declaration* was created as a result of the UN Conference on the Human Environment in 1972. The declaration states that humans have a shared responsibility to care for the Earth's resources and global environment. Some of the principles that the Stockholm Declaration outlines include

- That humans have a fundamental right to freedom, equality, and healthy living conditions
- That natural resources must be protected to benefit humans now and in the future
- That humans have a shared responsibility to maintain and manage the Earth's resources, wildlife, and habitat

Although the Stockholm Declaration doesn't bind any nation to specific action, it does outline that nations will work together in international cooperation to maintain a healthy, pollution-free global environment for the good of all humankind. As a result of this declaration, the UN created the UN Environment Program (UNEP) to help nations develop and implement more sustainable practices.

Montreal Protocol

The *Montreal Protocol* is an agreement signed by all UN members concerning the role humans play in the destruction of the stratospheric ozone layer (see Chapter 15 for details). Prior to current debates about climate change and global warming, every nation in the UN agreed that humans needed to halt ozone damage by air pollutants. After being signed in 1987 and put into action in 1989, the Montreal Protocol has been successful in phasing out the use of dangerous CFC aerosol chemicals and preventing further damage to the ozone layer.

The Montreal Protocol is recognized as an example of what the international community can achieve when they work together toward a common goal concerning Earth's shared resources.

UN Framework Convention on Climate Change

The *UN Framework Convention on Climate Change* (UNFCCC) is an agreement that came out of a large meeting in Rio de Janeiro in 1992 called the *Earth Summit.* Nations that were interested in understanding global climate change and reducing the impact of human activities on the world global climate system signed this agreement, which laid the groundwork for later agreements like the Kyoto Protocol (keep reading to find out details on this) and other protocol created as updates to the UNFCCC.

Kyoto Protocol

The *Kyoto Protocol* of 1997 is probably the most well-known of the UNFCCC updates concerning climate change. At a meeting in Kyoto, Japan, many nations agreed to set limits on carbon dioxide and other greenhouse gas emissions. The nations that signed the Kyoto Protocol agreed to a binding contract to limit their greenhouse gas emissions. Nations still in the process of industrial development, such as China and India, weren't required to sign the treaty, while other nations, such as the U.S., chose not to sign it. A few nations have succeeded in reducing their greenhouse gas emissions in line with Kyoto guidelines, but worldwide, greenhouse gas emissions haven't decreased significantly. Since the Kyoto Protocol, other meetings have been held in Copenhagen and Cancun with less success than the meeting in Kyoto, and the need for international cooperation to mitigate climate change continues.

Across the U.S., more than 500 individual cities, including New York, Los Angeles, Chicago, and Seattle, have taken it upon themselves to meet the emissions standards outlined by the Kyoto Protocol, regardless of whether the U.S. as an entire nation agrees to those standards. Their approaches vary, but each city seeks to reduce its greenhouse gas emissions by expanding public transportation and alternative energy use and encouraging more sustainable development to reduce urban sprawl.

Other international agreements

Some other international agreements concerning the environment include the following:

  - ✔ **Convention on the Conservation of Migratory Species of Wild Animals (CMS):** To conserve and protect species that migrate through the cooperation of multiple governments across the species' migration range

- ✔ **Basel Convention on the Control of Transboundary Movements of Hazardous Wastes and their Disposal (or the Basel Convention):** To protect the environment and human health from the dangers of improperly disposed hazardous waste

- ✔ **UN Convention to Combat Desertification (CCD):** To protect regions of the world susceptible to desertification through international partnerships, mitigation, and sustainable development

- ✔ **Convention Concerning the Protection of the World Cultural and Natural Heritage (or the World Heritage Convention):** To protect sites of cultural importance (natural or manmade) around the world from destruction

- ✔ **UN Convention on the Law of the Sea (UNCLOS):** To establish guidelines for international sharing of oceans and marine resources

From Sea to Shining Sea: Environmental Protection in the U.S.

At the end of the 1960s and throughout the 1970s, the U.S. passed a number of important laws on environmental issues, including pollution, species protection, and natural resource management. In this section, I describe the agencies responsible for protecting the environment and the most important U.S. environmental laws.

Overseeing environmental regulations

Multiple agencies in the U.S. play a role in protecting the environment and managing environmental resources. I list a few of the biggest ones here:

- ✔ **Environmental Protection Agency (EPA):** The EPA is the most well-known U.S. agency with an environmental focus. The EPA oversees most governmental science, research, education, and policy concerning the environment. It works with the Department of Justice and state-level agencies to enforce policy and regulations concerning environmental protection.

- ✔ **National Oceanic and Atmospheric Administration (NOAA):** As its name implies, NOAA focuses on the ocean and atmosphere. It's largely a research organization within the Department of Commerce that seeks to improve understanding of oceanic and atmospheric resources. NOAA plays a role in managing coastal and ocean resources, studying weather and climate, and collecting and interpreting huge amounts of data on subjects like global climate change, weather patterns, and oceanic ecosystems.

✔ **Fish and Wildlife Service (FWS):** The U.S. Fish and Wildlife Service in the Department of the Interior oversees the conservation of natural resources, including fish, animals, plants, and habitat, to meet the needs of the American people.

✔ **Occupational Safety and Health Administration (OSHA):** OSHA is an agency within the Department of Labor that oversees the health and safety of workers in the U.S. It plays a large role in limiting workers' exposure to toxic chemicals and other pollutants or unhealthy conditions in the workplace.

✔ **Department of Energy (DOE):** The DOE was created in 1977 to improve the energy security of the U.S. To fulfill this role, the DOE does research into both energy innovation and environmentally friendly energy resources. The Office of Environmental Management within the DOE oversees the environmental safety and cleanup of hazardous energy waste (such as nuclear waste).

Legislating for the environment

Until the 1960s and 1970s, the U.S. government did very little to protect natural resources and the environment. But in the late 1960s and throughout the 1970s, Congress passed multiple laws that shaped the legal landscape of environmental protection and natural resource conservation today.

National Environmental Policy Act of 1969

The *National Environmental Policy Act of 1969,* also known as NEPA, does three important things:

✔ Describes environmental policies and goals on a national scale

✔ Requires that federal agencies consider environmental outcomes and effects in their decision making

✔ Creates the Council on Environmental Quality to advise the president on environmental matters

NEPA asks federal agencies to complete an *environmental impact statement,* or EIS, to study in detail how a project will affect the environment and to explore possible alternatives that would be less harmful. NEPA doesn't block any federally funded project, regardless of its negative environmental impacts. Rather, NEPA simply requires that agencies make public the likely and potential outcomes of their projects so that U.S. taxpayers are informed about the environmental impacts. The fact that agencies have to inform the public is often enough to halt or postpone projects of grave environmental damage thanks to negative public opinion.

Clean Air Act of 1970 and 1990

The *Clean Air Act of 1970* (CAA) established the first set of standardized rules concerning air pollution on a national scale. Its overall goal is to improve air quality for the health of U.S. citizens. The CAA outlines guidelines for identifying *conventional* or *criteria* pollutants (the major pollutants; see Chapter 15) and regulating air pollution from industry and automobiles. Updated with amendments in 1990, the current Clean Air Act addresses the following issues:

- Cutting emissions of the most common, or criteria, air pollutants, including carbon monoxide, particulates, and lead

- Setting emissions standards for motor vehicles

- Controlling the release of toxic air pollutants

- Targeting sulfur dioxide and nitrogen oxide emissions to reduce acid rain

- Keeping the air in national parks clean and clear

- Reducing pollutants that cause ozone damage (see Chapter 15 for details on the ozone)

Clean Water Act of 1972

The *Clean Water Act of 1972* (CWA) and its amendments in 1977 regulate how pollutants are released into waterways in the U.S. The CWA sets standards for acceptable levels of certain pollutants in surface waters (though it doesn't cover groundwater contamination) and requires industries that release point source pollution (pollution with a clear origination) into waterways to get a permit first. Unfortunately, the CWA guidelines and permit procedures don't apply to nonpoint source water pollutants, which don't have a clear origination and are still a major problem (see Chapter 16 for more on water pollution).

The water quality standards set up by the CWA are viewed as some of the most effective environmental policy ever enacted. Water quality across the nation improved dramatically following the passing of the Clean Water Act. Prior to this act, there was no limit to what type of pollutants and how much waste industries could dump into the nation's lakes, rivers, and wetlands.

Endangered Species Act of 1973

The *Endangered Species Act of 1973* protects species from extinction and authorizes the U.S. Fish and Wildlife Service to create and implement recovery plans for endangered species. The act allows for protection of both individual species and the areas of habitat that endangered species depend on. Although many species listed as endangered have become extinct, the ESA has helped some species, including the bald eagle, peregrine falcon, grey whale, grizzly, grey wolf, and whooping crane, recover to levels healthy enough that they could be removed from the endangered species list.

Safe Drinking Water Act of 1974

While the Clean Water Act of 1970 regulates pollution in surface waters, the *Safe Drinking Water Act of 1974* (SDWA) targets waters above or below ground that are specifically designated for human drinking water. This act implements standards of health for drinking water based on peer-reviewed science. Amendments to the SDWA include requiring that public water supplies be constructed of lead-free pipes to reduce lead contamination and regulating the levels of bacteria in airline drinking water supplies. The act doesn't apply to bottled drinking water, which falls under the Food and Drug Administration's regulations.

Resource Conservation and Recovery Act of 1976

Congress created the *Resource Conservation and Recovery Act of 1976* (RCRA) as an amendment to the previously existing Solid Waste Disposal Act of 1965 to further regulate the disposal of waste. Most importantly, this act regulates hazardous waste from "cradle to grave," which means it includes all phases in the life cycle of hazardous waste from production to transportation to storage to final disposal. Other goals of the RCRA include reducing waste through recycling and keeping natural resources from being contaminated by waste.

National Forest Management Act of 1976

The *National Forest Management Act of 1976* requires that forest managers approach resource management from a more interdisciplinary perspective and consider the effects of timber removal on ecosystems. The act also asks forest managers to create resource management plans that consider other uses for the forest resources (such as recreation, mining, and ecosystem services) and approach logging in more sustainable ways.

Surface Mining Control and Reclamation Act of 1977

The *Surface Mining Control and Reclamation Act of 1977* regulates coal mining in the U.S. and creates a fund to pay for the cleanup and restoration of land used for mining. In particular, the act targets strip mining for coal, which dramatically changes the shapes of the Earth's surface, by requiring that surface contours be returned to their original conditions to reduce the impacts of mining on the landscape. This act also created a federal agency to oversee consistency in coal mining regulations across state boundaries.

Comprehensive Environmental Response, Compensation, and Liability Act of 1980

The *Comprehensive Environmental Response, Compensation, and Liability Act of 1980,* also known as CERCLA, established the Superfund to finance the cleanup of toxic waste at sites across the nation. (I describe the Superfund and some of these toxic waste sites in Chapter 18.) In 1994, Congress amended this act to require that the parties responsible for the hazardous or toxic waste mess share responsibility for cleanup costs.

Part VI

The Part of Tens

In this part . . .

Learning about the environment is only one part of living more sustainably. In this part, I list ten simple ways for you to shift toward a more sustainable lifestyle. I also describe ten examples of how the tragedy of the commons has played out in real life — including ways that communities and nations are cooperating to restore the damage that has already been done.

Finally, I explain how you can put your environmental science knowledge to work with a list of ten careers that involve environmental science in some way.

Chapter 22

Ten Ways to Live Sustainably

*E*nvironmental science is all about finding ways to live more sustainably, which means using resources today in a way that maintains their supplies for the future. Environmental sustainability doesn't mean living without luxuries but rather being aware of your resource consumption and reducing unnecessary waste.

In this chapter, I describe ten simple ways to live more sustainably. These strategies aren't simply ways to be "green," a common phrase used in marketing these days to capture the attention of people concerned about the environment. These are ten changes — some minor, some major — in how you think about your place in the world and how you use the Earth's resources. Don't be afraid to start small; even the simplest change in your daily life can accumulate into something meaningful on a global scale.

With each change you make, you show the people around you that living sustainably isn't about going without or about hugging trees but about maintaining Earth's ecosystem so that human beings, and other living things, continue to thrive.

Reducing Household Energy Use

As I describe in Chapter 14, energy conservation is itself a source of energy. Conserving energy has long been considered a great way to save money on monthly utility bills, and it has only become more important in the last few decades as political struggles and climate change have highlighted the high environmental and societal costs of fossil fuels.

Here are several simple ways to reduce your household energy use:

- ✔ Turn off appliances and lights that you're not using.
- ✔ Install energy-efficient appliances.
- ✔ Use a programmable thermostat that lowers the temperature (or raises it, depending on the season) when you're not home.
- ✔ Set your thermostat lower than usual in the winter and bundle up.
- ✔ Open windows to allow a breeze instead of turning on the air conditioning.
- ✔ Hang clothes to dry instead of using the dryer.
- ✔ Use an electric teakettle rather than a stovetop kettle to boil water. (Doing so reduces waste that occurs in the energy transfer between the stovetop and the kettle. Plus, electric teakettles are more efficient and heat up much faster.)
- ✔ Replace incandescent light bulbs with compact fluorescent bulbs (CFLs).

Keep in mind that you pay for both the energy you use and the energy you waste. By reducing your energy use, you save money and energy so that fewer natural resources are necessary to fuel your daily life.

Eating Locally

A powerful way to live more sustainably is to eat locally. The convenience of supermarkets has changed how people think about food. You can stroll through aisles stocked with fruits, vegetables, and other products from all over the world any time of year. But these products consume huge amounts of fossil fuel energy to get from those global locations to your corner supermarket.

Many cities in the U.S. have begun to support farmer's markets that bring regional food producers together with consumers looking to live more sustainably. Of course, eating local may mean that you don't have year-round access to, say, bananas, but you'll discover which fruits and vegetables are in season in your region and probably be surprised at how delicious locally grown and ripened produce can be.

Moving toward sustainable living doesn't need to be an all-or-nothing choice; don't throw in the towel completely just because you feel you can't be "perfect." Even if you buy your vegetables locally but still savor those bananas from South America, you're making a positive difference. Depending *more* (if not *completely*) on locally available foods still reduces the amount of energy used in food transportation and supports your local food-producing economy.

Disposing with Disposables

The last 50 years have seen a revolution in the plastics industry. Previous generations didn't dream of single-use razors, forks, cups, bags, and food storage containers, but these days, you can find a plastic version of almost any object and then throw that object away after you use it. Disposable items come in bulk at low prices, but the real price is paid by the environment and human health: Large landfills, trash on roadsides, and toxins from disposable products are found almost everywhere.

Many of the environmental health issues today stem from toxins released into the environment by trash. Even trash that's properly disposed of, such as that in a landfill, requires careful monitoring to ensure that dangerous chemicals don't enter the surrounding environment. You've certainly heard the common tagline "reduce, reuse, recycle," and these three practices certainly help reduce the amount of waste entering landfills and the environment. But consider a fourth *r: rethinking* your dependence on disposable products.

When you make a purchase, consider the item's life expectancy: How long can the item be used? Will it have more than one use? When you're done with it, will it end up in the trash? Start investing in reusable products for the items you most often throw away, such as the following:

- A coffee mug, travel cup, or thermos to carry hot drinks with you
- A razor with refillable blades
- A refillable lighter
- Fabric shopping bags for groceries and other shopping trips
- Sturdy, lightweight, washable containers to pack your lunch in
- Cloth napkins and dishtowels to replace disposable napkins and paper towels
- Alternative gift wrap, such as reusable cloth bags or other containers, rather than ribbon and traditional wrapping paper that end up in the trash

When you do require disposable products, make more sustainable choices by thinking about what materials went into making the product, whether it can be recycled, and whether it biodegrades. Use paper or *bioplastics* (plastic made of corn or other recycled plant matter) when possible, and always look for plastics made of *post-consumer materials* that are recyclable.

Planting Seeds

Try growing your own food. No need to buy a tractor; start small. Simply plant a few seeds in a corner of your yard or in a container on your porch or windowsill. You don't need acres; a few square feet on a patio, along the driveway, or in a window box can provide enough space to grow edible herbs, fruits, and vegetables. Growing food plants in containers is a great way to control the health of the soil and provide or change proper light and water conditions. And because you're the farmer, you can control exactly what goes into the soil and can avoid using toxic pesticides.

If you do have lots of space and a love for gardening, try adding some extra rows to your vegetable garden and then donate the extra crops to a local food bank or charity organization. Some places take extra home-grown crops so that folks who don't have resources to grow their own food still have access to healthy, local, inexpensive produce.

Recycling

Recycle as much as possible! If your neighborhood or apartment complex doesn't offer recycling pickup, either find a drop-off location or request the curbside service. Recycling services may cost a small fee just like garbage collection, but consider the value of recycling: It reduces trash and conserves natural resources. Paper, plastic, glass, aluminum, and tin are commonly accepted recyclables that can be used to make new consumer products. If you see products labeled *post-consumer,* you know they've been made from recycled materials. Buying these items lets companies know that recycling is the way to go!

For other items, such as CFLs, batteries, cellphones, and electronics, find out where you can drop off these items with an appropriate recycler. Be sure to ask electronics recyclers where these materials go for recycling and avoid companies that ship electronic waste overseas for the unregulated "recycling" and salvage operations I discuss in Chapter 18. Goodwill Industries International (www.goodwill.org) is one place that accepts electronics for responsible recycling.

Most cities and towns also have recycling centers that collect and recycle or properly dispose of hazardous household products, such as paint, oil, and other items that should be kept out of landfills. Look on the Internet (earth911.com) or in the phone book for one near you so that you can finally get rid of those old cans of paint collecting dust in your garage or basement.

Reselling and Donating Items

Items that you no longer need can get an extended life through resale and donation. Although you may no longer have need or use for an item, someone else may. By extending the life of any product, you help reduce dependence on disposable or cheaply made, single-use products that end up in landfills.

Try reselling clothing and children's things through a secondhand or consignment retailer or consider donating them to a nonprofit resale organization (such as Goodwill) or charity organization (such as the Salvation Army or American Cancer Society) that will redistribute them to those in need. Other options for reselling items today include websites such as www.craigslist.org or www.ebay.com, where buyers and sellers of all different items meet.

Drinking from the Tap

In recent years, dependence on bottled water has added more than a million tons of plastic to the waste stream every year. One reason people rely on bottled water is because they believe it's safer and better tasting than tap water. But most municipal water supplies in the U.S. provide safe, clean, fresh water (and many bottled waters are just bottled from city water supplies anyway).

If the flavor of your tap water is what sends you reaching for a bottle, consider the one-time investment in a filtration system; you'll save money in the long run and help the environment to boot. If the convenience of bottled water is what attracts you, purchase refillable bottles and keep one in your fridge, one in your car, and one at the office. Encourage your employer to install filters and offer glasses or reusable bottles at work, too.

Saving Water

An easy way to live more sustainably is to conserve household water use. When shopping for new appliances, look for information about how they can save water (and energy). For example, consider installing water-efficient toilets or dual-flush toilets that let you choose whether to use a full flush (for solid waste) or half-flush (for liquid waste). Also, newer clothes washers can automatically sense the smallest level of water needed for each load.

Smaller changes, such as switching to water-saving shower heads and adding aerators to your sink faucets, are also effective ways to significantly reduce household water use. If new appliances are beyond your budget, try these simple ways to reduce water use:

- ✔ Turn off faucets while brushing your teeth and washing your hands.
- ✔ Take short showers.
- ✔ Wait until you have a full load to run the dishwasher or clothes washer.

Outdoor water use is one place where households often waste water. To conserve water outdoors, use landscaping adapted to your local environment. Native plants require less overall care and often survive on less watering than plants adapted to other regions. When buying plants, look for drought-tolerant species and varieties and be sure to plant them in proper soil and sun conditions to reduce their need for excess watering. Also, set up sprinkler systems so they don't water the sidewalk, the driveway, and other paved, impermeable surfaces. Setting sprinkler systems on a timer to water in the morning or evening during hot summer weather helps reduce evaporation so that the water being used goes directly to your plants.

If you live in a region that gets seasonal rainfall, look into adding *rain barrels* at the end of each gutter drainpipe. These large barrels capture water that washes off your roof. When it's time to water the lawn or garden, you simply use the rainwater that you captured for free instead of running the faucet.

Relying Less on Your Car

Using fossil fuels to support one person in each car on the road is clearly no longer sustainable. Higher fuel prices, increasing political and economic conflict over access to fuel resources, and concerns about the effect of increasing carbon dioxide in the atmosphere have led many people to look for alternatives to using their cars.

Investigate mass transit options in your town or city, such as a bus system, a light rail train system, or carpool and vanpool services for commuters. When traveling close to home, walk or ride your bike. Some people have found that they can live without a car entirely by depending on public transit options and the occasional use of shared car services such as Zipcar (www.zipcar.com).

If you really can't live without your car, carefully plan when and where you drive. Save errand running for one day a week, and look at your list before leaving to organize the order of your stops so that you're driving as little as possible. Perhaps you can drive downtown and then run your errands on foot or join a neighbor to run errands together.

Many American towns and cities are just getting the hang of sustainable development and have begun to build neighborhoods within walking distance of commercial centers. Let your city planners and policymakers know that sustainable development is something you want to see more of in your town!

Purchasing Fair-Trade Products

When you purchase items that are imported from all over the world — particularly coffee, cocoa, sugar, tea, chocolate, and fruit — look for the fair-trade certification. The *fair-trade* designation tells you that these items were grown using sustainable methods of agriculture and that local people are receiving fair prices for the goods they produce.

Items that don't have the fair-trade certification may have been produced unsustainably and may be the product of exploitative labor practices that don't benefit the local people.

When you purchase fair-trade products, you're making the statement that all people deserve to participate in fair business practices that support sustainable methods of resource use.

Chapter 23

Ten Real-Life Examples of the Tragedy of the Commons

*T*he phrase *tragedy of the commons* was first described by biologist Garrett Hardin in 1968. In a scientific paper, he described how shared environmental resources are overused and eventually depleted. He compared shared resources to a common grazing pasture; in this scenario, everyone with rights to the pasture grazes as many animals as possible, acting in self-interest for the greatest short-term personal gain. Eventually, they use up all the grass in the pasture; the shared resource is depleted and no longer useful.

In this chapter, I offer ten examples of the tragedy of the commons. Each is an example of resources being depleted; in some cases, lack of planning has destroyed the resources beyond repair, but other cases show that cooperative agreement and sustainable management can help avoid the tragedy of the commons.

Grand Banks Fisheries

The Grand Banks are fishing grounds off the coast of Newfoundland in the northwestern Atlantic Ocean. For centuries, explorers and fishermen described this region as home to an endless supply of cod fish. This prospect attracted fishermen from all over the north Atlantic and provided the basis for a thriving fishery industry in Canada.

In the 1960s and 1970s, advances in fishing technology allowed huge catches of cod; the maximum reported one season was 800,000 tons. Following a few dramatically large seasons, the fish populations dropped, forcing Canadian fisherman to sail farther to maintain large catch sizes each season. Before long, they were dragging large seafloor *trawlers* along the bottom of the ocean to catch as many of the remaining fish as possible. These methods destroyed not only the fish populations but also the entire seafloor ecosystem of the Grand Banks.

By the 1990s, cod populations were so low that the Grand Banks fishing industry collapsed. It was too late for regulation and management; the cod stocks had been irreparably damaged. Since then, the cod populations have remained low, and some scientists doubt the Grand Banks ecosystem will ever recover.

Bluefin Tuna

Currently the bluefin tuna populations in the Atlantic Ocean and Mediterranean face a similar fate as that of the Grand Banks cod (see the preceding section). In the 1960s, fishermen realized the tuna populations were in danger, and an International Convention for the Conservation of Atlantic Tuna (ICCAT) formed in an effort to manage fish harvesting more sustainably. Unfortunately, not every nation is a member of the ICCAT or follows the convention's guidelines. Instead, many nations continue to seek profit from large bluefin tuna catches every year without regard for conservation. Bluefin tuna have already been fished to extinction in the Black Sea and Caspian Sea, and perhaps the Atlantic bluefin tuna will follow.

Passenger Pigeons

When Europeans arrived in North America, passenger pigeons migrated across the sky in huge numbers; some flocks reportedly darkened the sky for up to 14 hours as millions of birds flew by. Scientists estimate that about 4 billion passenger pigeons lived in North America in 1800.

As European populations spread farther into the continent, they began to clear the forests that passenger pigeons inhabited (destroying the birds' habitat) and eventually began to hunt the pigeons for food. In the mid-1800s, they caught massive numbers of pigeons in nets and sold the birds in cities as a food resource. The tendency of the pigeons to travel and nest in large flocks made them easy targets for hunters seeking profits. At the time, no limits dictated how many pigeons hunters could catch and kill each year or how to sustain the pigeon populations for the following years.

By 1870, nearly all the passenger pigeons had been killed; hunting limits were enacted in the 1890s, but by that time, the passenger pigeon population couldn't recover. The last known passenger pigeon (held in captivity at a zoo) died in 1914, completing the extinction of a species because of unsustainable hunting practices.

Ocean Garbage Gyres

The ocean is an excellent example of a shared resource that can easily be abused and degraded because it's shared by many different nations. No single authority has the power to pass laws that protect the entire ocean. Instead, each nation manages and protects the ocean resources along its coastlines, leaving the shared common space beyond any particular jurisdiction vulnerable to pollution.

Throughout the world's oceans, garbage has begun to accumulate in the center of circular currents, or *gyres* (check out Chapter 18 for details). These giant patches of ocean garbage occur because many different countries allow solid waste to enter the oceans from land or ships.

Destruction of ocean ecosystems because of garbage, especially plastic pollutants, is likely to affect every person on the planet as these pollutants cycle through the food chain. Unless the international community comes to an agreement to protect and preserve the oceans from garbage pollution, this issue may become the most dramatic global example of a tragedy of the commons.

Earth's Atmosphere

Much like the oceans, Earth's atmosphere is a resource that everyone on the planet uses and abuses. Air pollution and greenhouse gases from various industries and transportation increasingly damage this valuable, shared resource.

As an example of a tragedy of the commons, the atmosphere offers some hope for a solution: More than once, international agreements have recognized the importance of taking care of the atmosphere. One example is the Montreal Protocol, which slowed ozone depletion (see Chapter 15 for details). Another is the Kyoto Protocol, which attempted to bring nations together in reducing greenhouse gas emissions and slowing global climate warming (see Chapter 19). In the Kyoto meetings and the agreement that came out of them, multiple nations recognized that everyone had an interest in preserving this common resource for the future and agreed to look beyond short-term gain and immediate self-interest to a sustainable future.

Unfortunately, not every nation signed the Kyoto Protocol agreement, including the U.S., and even some of those who did sign the agreement continue to degrade the shared resource of Earth's atmosphere with excessive carbon emissions and other air pollutants.

Gulf of Mexico Dead Zone

Thousands of farms are located along the Mississippi River and its tributaries through the central U.S. As water washes into the river after a heavy rain, it brings with it nutrients from fertilizers added to farmland. These materials flow downriver and eventually enter the Gulf of Mexico, where they create conditions for a *dead zone* — a region of the ecosystem that can't support any living creatures. (I explain exactly how nutrient pollution creates dead zones in Chapter 16.)

The Gulf of Mexico has a dead zone because everyone along the Mississippi River shares the waterway without considering how each small contribution of nutrient and chemical pollution will have dramatic results. Each farmer has a right to treat his crops with fertilizers to boost production, but so many people use these chemicals that by the time the river empties into the Gulf, the chemical load is enough to disrupt the ecosystem.

The Gulf of Mexico isn't the only region that experiences dead zones. All over the world, dead zones occur wherever rivers flow through agricultural lands before entering a larger body of water (such as a sea or gulf).

Traffic Congestion

Traffic congestion is a case of the tragedy of the commons that plays out every day, twice a day, during rush hour. Public roads, constructed and maintained by the government, are an excellent example of common property shared by many people. Each of these people has his or her own interest in mind — typically, how to get to work as quickly and easily as possible. But when everyone decides that public roads are the best way to meet traveling needs, the roads jam up and slow down overall traffic movement, filling the air with pollutants from idling cars.

Turning public roads into private roads or toll roads creates a different scenario. With a toll to pay (especially if the toll is higher during peak-use hours such as rush hour), drivers may consider a less-direct route or choose to drive to work at a different time. All these changes help reduce the effects of congestion on free, public roads.

Groundwater in Los Angeles

Landowners around Los Angeles each have rights to use the water pumped up from wells on their land. This water is part of a regional groundwater aquifer, so each landowner is ultimately pulling water from the same pool. As the city grew in the 1930s and 1940s, the amount of water drawn from the underground aquifer increased each year to meet the needs of the growing population.

Eventually, residents drew so much water from the aquifer that the supply reached levels that left the aquifer vulnerable to saltwater intrusion from the nearby Pacific Ocean. (Flip to Chapter 9 for details on how saltwater intrusion affects aquifers.) Facing potential water shortages and possible destruction of the renewable water resource they depended on, the water users created a voluntary organization to discuss how to manage and conserve the groundwater for the future.

Unregulated Logging

The tropical rainforests are a common resource that everyone in the world benefits from, even though you may not realize it. (I explain all the benefits of biodiversity — most of which are found in these forests — in Chapter 12.) In some parts of the world, such as South America, vast expanses of dense rainforests aren't governed or owned in a way that allows effective management for resource extraction. Timber producers are driven to remove as much timber as possible as cheaply as possible. The result is that logging irreparably damages acres of rainforest each year.

Although some laws protect these forests from destructive logging practices, illegal logging continues — particularly along boundaries between countries, where the laws may be different on each side of the border. As I describe in Chapter 10, less-disruptive methods of timber production exist, but they often require more time and money in their effort to not damage the ecosystem.

Population Growth

Some scientists consider the exponential growth of the human populations to be an example of a tragedy of the commons. In this case, the common resource is the planet Earth and all its shared resources, including (but not limited to) the ones I describe in the preceding sections. As I write this book, the world's population has reached a whopping 7 billion individuals. (***Remember:*** This number is an estimate — probably an underestimate — because no international census bureau exists to count world population numbers.)

Examining population growth as a tragedy of the commons illustrates that the depletion of common resources isn't always the result of greed. Just by existing, each person uses water, air, land, and food resources; splitting those resources among 7 billion people (and counting) tends to stretch them pretty thin.

One way to tackle this resource strain is to encourage each person to take responsibility for how many resources he or she uses. This accountability means conserving water, reducing air pollution, supporting local farms or planting a vegetable garden of your own, and practicing other methods of reducing your individual ecofootprint (see Chapter 5 for an explanation of ecofootprints and Chapter 22 for some tips on living sustainably).

Chapter 24

Ten Careers in Environmental Science

In This Chapter

▶ Selling sustainability, conserving resources, and defending the environment

▶ Teaching others about the environment

▶ Incorporating sustainable ideas into every career

*O*ne of the best outcomes of studying environmental science is realizing that you can play a role in creating a cleaner, more sustainable world to live in. Even if you don't plan to pursue a career in environmental science itself, many students find that whatever their career choice, they can incorporate consideration of the ecosystem, energy use, pollution, or human health into their daily work. This chapter describes a few of the options for careers in environmental science and some ways to incorporate your environmental science education into whatever you choose to do.

Marketing Sustainability

As more people seek to live more sustainably, companies that sell products will want to meet that need, and they'll need marketing professionals to help them reach the public. In marketing, strategies for selling products to the people who want or need them intersect with a business's profit goals; your job could be to align the two. Already green industries have seen extensive growth, and products that are more environmentally friendly and sustainable have become the norm in many businesses and households.

Selling people products based on how environmentally friendly they are isn't the only way to use marketing in the context of environmental science. You can also sell the idea of sustainability to the companies that manufacture and produce consumer products. For example, you could find ways to convince commercial industries that sustainability is part of a long-term business plan and then develop transitional paths from current unsustainable practices toward sustainable practices for the future.

Restoring Natural Landscapes

Where efforts to conserve or preserve ecosystems have failed, the field of restoration ecology comes to the rescue. Restoration ecologists find places that have been damaged and seek to restore them to their natural state. For example, as a restoration ecologist, you might change vacant lots filled with weeds into natural habitats for local plants and animals or replant native plant species along highways and roads where invasive species have taken over.

Restoration ecologists are at work in every city and town as are landscape designers, engineers, botanists, greenhouse technicians, and everyday gardeners. Some work on developing educational programs about habitat restoration, while others organize volunteers to help plant trees.

Spreading the Word and Educating Others

One of the most important jobs in environmental science is helping people understand the issues related to pollution, environmental damage, and ecosystem health. If you're interested in spreading the word about environmental issues, you have numerous career options available, including journalism, photography, writing, and public speaking.

Think of what led you to study environmental science — a photograph, an article, or a story about the environment that brought your attention to issues of sustainability. Whether you're interested in careers in communication (visual, print, or broadcast journalism) or you're more artistically driven (photography and writing), you can help spread awareness about the environmental issues you feel most strongly about.

Many students who study environmental science find they want to teach others about how ecosystems function and how environmentally damaging practices affect each and every living thing. Careers in environmental education range from teaching in a classroom to spending your days outdoors. Many national parks, wildlife reserves, zoos, parks, and recreation areas employ educators to take visitors on tours. These tours combine the experience of the natural beauty with an understanding of ecosystems and environmental health and sustainability.

Continuing Study and Research

Although you may think that almost everything about the natural world and the way living things interact is already known, that's far from the truth! Every day scientists studying ecology, biology, chemistry, toxicology, geology, atmospheric science, oceanography, and many other sciences add to the knowledge used to create more sustainable solutions to environmental problems around the world.

A scientific research career in any area of study has some environmental applications. Scientists working in the field, in the lab, and in the library drive forward scientific understanding of ecosystem interactions and the effects of environmental pollution. While scientific progress appears to move forward one tiny step at a time, the combination of all these tiny steps is what adds up to real, visible progress in the accumulation of scientific knowledge.

Defending the Environment

If you're interested in enforcing regulations and making sure that anyone who doesn't comply with environmental health and safety laws is held responsible, environmental law may be the career path for you.

Practicing environmental law gives you an opportunity to combine legal knowledge and research with the environmental values of sustainability and responsible stewardship of the Earth. Some aspects of environmental law include conservation, pollution control, remediation, resource management, and urban planning and infrastructure.

Assessing Risk

A career in risk assessment involves doing scientific sampling, analysis, and study of environmental conditions to determine what ecosystem or human health risks are present in a given situation. Private companies and government agencies often hire risk assessors to analyze how current practices are affecting the environment. For example, assessors may study the current levels of pollutants in a lake and describe how those pollutant levels are likely to affect the lake's ecosystem or the people who live nearby.

Based on the risk assessment, companies and agencies will develop new approaches or reformulate existing practices to comply with laws or to reduce their environmental impacts.

Analyzing Policy

As policymakers seek to create a more sustainable world through laws and regulations, they need the help of someone in the know. That's where you'd come in if you were an environmental policy analyst. Environmental policy analysts help create new policy, analyze existing policies, and suggest improvements. They examine the social, environmental, and economic implications and outcomes of government (or private corporation) environmental laws, and they play an important role in developing more sustainable practices for both government agencies and independent corporations.

Engineering Solutions

A little careful planning goes a long way toward fixing (or avoiding) many environmental problems. In recent decades, environmental engineers have played an important role in creating infrastructure that helps keep the environment clean and safe. Some of the first environmental engineers built the Roman aqueducts — bringing fresh water to the Roman people. Nowadays environmental engineers construct everything from sewage systems and hydropower dams to cleaner industrial air and waste systems.

Conserving Farm and Ranch Land

Many private consulting firms and government agencies have positions for conservation scientists, who act as important consultants for farmers, ranchers, and resource managers seeking to conserve or use their land and resources in sustainable ways. As a conservation scientist, you could help a farmer or rancher transition away from environmentally damaging practices to sustainable long-term approaches that maintain their livelihood as well as the health of the ecosystem they depend on.

Conservation scientists are concerned with preserving natural landscapes and finding ways for human resource use and natural ecosystems to coexist. Using sustainable approaches to farming, ranching, logging, and other types of land use means that these resources will be available for human use long into the future.

Advising Investment

If your career interests lie more in finance, you can still keep the environment in mind. Many investors today are concerned with where their money is being spent but don't necessarily have time to research every company in their portfolio's mutual funds. An investment advisor with some experience, education, and interest in the environment can play an important role in helping people choose the most sustainable companies to invest in.

Index

• T •

• U •

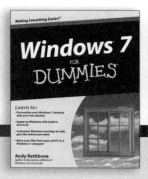

Math & Science

Algebra I For Dummies,
2nd Edition
978-0-470-55964-2

Biology For Dummies,
2nd Edition
978-0-470-59875-7

Chemistry For Dummies,
2nd Edition
978-1-1180-0730-3

Geometry For Dummies,
2nd Edition
978-0-470-08946-0

Pre-Algebra Essentials
For Dummies
978-0-470-61838-7

Microsoft Office

Excel 2010 For Dummies
978-0-470-48953-6

Office 2010 All-in-One
For Dummies
978-0-470-49748-7

Office 2011 for Mac
For Dummies
978-0-470-87869-9

Word 2010
For Dummies
978-0-470-48772-3

Music

Guitar For Dummies,
2nd Edition
978-0-7645-9904-0

Clarinet For Dummies
978-0-470-58477-4

iPod & iTunes
For Dummies,
9th Edition
978-1-118-13060-5

Pets

Cats For Dummies,
2nd Edition
978-0-7645-5275-5

Dogs All-in One
For Dummies
978-0470-52978-2

Saltwater Aquariums
For Dummies
978-0-470-06805-2

Religion & Inspiration

The Bible For Dummies
978-0-7645-5296-0

Catholicism For Dummies,
2nd Edition
978-1-118-07778-8

Spirituality For Dummies,
2nd Edition
978-0-470-19142-2

Self-Help & Relationships

Happiness For Dummies
978-0-470-28171-0

Overcoming Anxiety
For Dummies,
2nd Edition
978-0-470-57441-6

Seniors

Crosswords For Seniors
For Dummies
978-0-470-49157-7

iPad 2 For Seniors
For Dummies, 3rd Edition
978-1-118-17678-8

Laptops & Tablets
For Seniors For Dummies,
2nd Edition
978-1-118-09596-6

Smartphones & Tablets

BlackBerry For Dummies,
5th Edition
978-1-118-10035-6

Droid X2 For Dummies
978-1-118-14864-8

HTC ThunderBolt
For Dummies
978-1-118-07601-9

MOTOROLA XOOM
For Dummies
978-1-118-08835-7

Sports

Basketball For Dummies,
3rd Edition
978-1-118-07374-2

Football For Dummies,
2nd Edition
978-1-118-01261-1

Golf For Dummies,
4th Edition
978-0-470-88279-5

Test Prep

ACT For Dummies,
5th Edition
978-1-118-01259-8

ASVAB For Dummies,
3rd Edition
978-0-470-63760-9

The GRE Test For
Dummies, 7th Edition
978-0-470-00919-2

Police Officer Exam
For Dummies
978-0-470-88724-0

Series 7 Exam
For Dummies
978-0-470-09932-2

Web Development

HTML, CSS, & XHTML
For Dummies, 7th Editio
978-0-470-91659-9

Drupal For Dummies,
2nd Edition
978-1-118-08348-2

Windows 7

Windows 7
For Dummies
978-0-470-49743-2

Windows 7
For Dummies,
Book + DVD Bundle
978-0-470-52398-8

Windows 7 All-in-One
For Dummies
978-0-470-48763-1

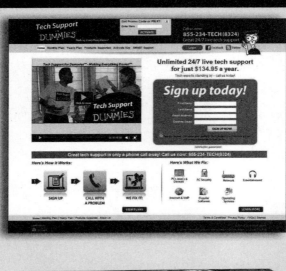